AF565115

PROGRESS IN COLLOID & POLYMER SCIENCE
Editors: H.-G. Kilian (Ulm) and G. Lagaly (Kiel)

---

Volume 92 (1993)

# Orientational Phenomena in Polymers

Guest Editors:
L. Myasnikova (St. Petersburg) and
V. A. Marikhin (St. Petersburg)

---

Steinkopff Verlag · Darmstadt
Springer-Verlag · New York

Die Deutsche Bibliothek – CIP-Einheitsaufnahme

**Orientational phenomena in polymers** / guest ed.: L. Myasnikova and V. A. Marikhin. – Darmstadt : Steinkopff ; New York : Springer, 1993
(Progress in colloid & polymer science ; Vol. 92)
ISBN 3-7985-0954-9 (Steinkopff) Gb.
ISBN 0-387-91453-6 (Springer) Gb.
NE: Mjasnikova, Ljuba [Hrsg.]: GT

ISBN 3-7985-0954-9 (FRG)
ISBN 0-387-91453-6 (USA)
ISSN 0340-255 X

This work is subject to copyright. All rights are reserved, whether the whole or part of the material is concerned, specifically those rights of translation, reprinting, reuse of illustrations, recitation, broadcasting, reproduction on microfilms or in other ways, and storage in data banks. Duplication of this publication or parts thereof is only permitted under the provisions of the German Copyright Law of September 9, 1965, in its version of June 24, 1985, and a copyright fee must always be paid. Violations fall under the prosecution act of the German Copyright Law.

© 1993 by Dr. Dietrich Steinkopff Verlag GmbH & Co. KG, Darmstadt.
Chemistry editor: Dr. Maria Magdalene Nabbe; English editor: James C. Willis; Production: Holger Frey, Bärbel Flauaus.

Printed in Germany.

The use of registered names, trademarks, etc. in this publication does not imply, even in the absence of specific statement, that such names are exempt from the relevant protective laws and regulations and therefore free for general use.

Type-Setting: Graphische Textverarbeitung, Hans Vilhard, D-64753 Brombachtal
Printing: Betz-Druck, D-64291 Darmstadt

# Preface

The 25th European Macromolecular Physics Conference on Orientational Phenomena in Polymers was held in St. Petersburg, Russia, July 6—10, 1992. It was organized by the Macromolecular Board of the European Physical Society, the Russian Academy of Science, the Ioffe Physico-Technical Institute and the Institute of Macromolecular Compounds. The excellent organization was managed by Professor Dr. V. A. Marikhin and Dr. L. Myasnikova. The Conference brought together scientist who are competent in the field of oriented macromolecular systems. Current achievements and newest developments were discussed while elucidating recent progress. It was possible to identify theoretical and experimental questions that should be tackled in the future so as to improve the methods of characterizing and processing chain alignment up to extremely large orientation.

The conference was attended by about 130 scientists from 46 universities, research centers, and industry. The lectures encompassed a wide range of studies dealing with the mechanism of orienting macromolecules by flow and magnetic fields, by solid phase deformation or epitaxial growth on oriented substrates. The resulting properties were discussed. The fine structure, nature and role of defects including fracture of oriented polymers were considered. New techniques of producing high performance polymer material were reported. Attention was given to the very fast developing field of electrical conductivity and polymers with nonlinear optical properties.

This volume contains part of the lectures presented at the conference.

The atmosphere and the spirit of the meeting was excellent, and to have directly encountered the way of life, the history, and the cultural of Russia was impressive and instructive.

L. Myasnikova,
V. A. Marikhin (St. Petersburg)

# Contents

**Preface** . . . . . . V

Ziabicki A: Orientation mechanisms in the development of high-performance fibers . . . . . . 1
Jasse B, Tassin JF, Monnerie L: Orientation and chain relaxation of amorphous polymers and compatible polymer blends . . . . . . 8
Bassett CD, Freedman AM: Lamellar morphologies in uniaxially-drawn banded spherulites of polyethylene . 23
Wittmann JC, Lotz B, Smith P: Formation of highly oriented films by epitaxial crystallization on polymeric substrates . . . . . . 32
Marikhin VA, Myasnikova LP: Structural basis of high-strength high-modulus polymers . . . . . . 39
Pertsev NA: Transformations of defect structure of polymers during drawing . . . . . . 52
Kilian HG, Knechtel W, Heise B, Zrinyi M: Orientation in networklike polymer systems. The role of extremum principles . . . . . . 60
Keller A, Kolnaar WH: Chain extension and orientation: Fundamentals and relevance to processing and products . . . . . . 81
Ward IM: New developments in the production of high modulus and high strength flexible polymers . . . . . . 103
Albrecht C, Lieser G, Wegner G: Lamellar morphology of polydiacetylene thin films and its correlation with chain lengths . . . . . . 111
van der Sanden MCM, Meijer HEH, Lemstra PJ: The ultimate toughness of polymers. The influence of network and microscopic structure . . . . . . 120
Gedde UW, Andersson H, Hellermark C, Jonsson H, Sahlén F, Hult A: Synthesis, characterization and relaxation of highly organized side-chain liquid crystalline polymers . . . . . . 129

**Author Index** . . . . . . 135

**Subject Index** . . . . . . 136

# Orientation mechanisms in the development of high-performance fibers

A. Ziabicki

Institute of Fundamental Technological Research, Polish Academy of Sciences, Warsaw, Poland

*Abstract:* Highly oriented polymer fibers can be manufactured in various ways involving different regimes of deformation. Two processes and two different mechanisms of orientation are discussed. The first is strain orientation in a plastic solid subjected to slow deformation. The example of industrial processes include cold-drawing, calendering, and solid-state extrusion. The other process is flow orientation in the fluid state (melt, solution, suspension). Melt- or wet-spinning are typical examples. The theory of orientation and stress in plastic-state and fluid-state processing is developed, and examples of industrial processes leading to high-performance fibers are discussed.

*Key words:* Molecular orientation — stress — orientation-stress characteristics — fibers — spinning — drawing — rotational diffusion coefficient — strain — strain rate

## Introduction

Polymer fibers with very high modulus and/or tenacity are widely used for reinforcement of composites, manufacturing of ropes, sails, geotextiles, and other products. The necessary structural feature is high degree of order: molecular orientation and crystallinity. All high-performance fibers: aramides (Kevlar, Technora), poly(phenyleno-benzo-thiazole) (PBT), poly(phenyleno-benzo-oxazole) (PBO), ultra-high-molecular-weight polyethylene (Dyneema, Spectra) are composed of highly oriented, linear macromolecules and exhibit unique mechanical properties.

High-performance fibers can be oriented in a variety of ways. Two classes of deformation regimes are used in industrial processes:

i) plastic deformation of a solid polymer
ii) flow of a fluid polymer.

The kinematics and dynamics of orientation are different in both classes, and different processes have to be chosen for different materials.

In the case of pseudo-plastic systems, orientation is controlled by strain in the system, $\varepsilon(t)$. Discussing fibers, we confine our considerations to uniaxial elongation with the non-linear measure of strain

$$\varepsilon(t) = \ln[L(t)/L_0] = \ln[R(t)] \quad (1)$$

for a discontinuous deformation of a sample with initial length $L_0$, or

$$\varepsilon(t) = \ln[V(x)/V_0] \quad (2)$$

for steady-state, continuous elongation. $V$ is axial velocity, $x$ — axial position in the deformed filament, and $R$ — draw ratio. Examples of plastic processing include cold drawing of fibers, solid-state extrusion, forging, calendering, etc.

Fluid-state processing, on the other hand, (fiber spinning, film casting, extrusion and injection) is controlled by strain rate, $\dot{\varepsilon}$, rather than the extent of strain, $\varepsilon$. In a fluid composed of highly mobile macromolecules, orientation is a result of competition between the orienting effect of the flow field, and disorienting effect of Brownian motions [1]. In the case of fibers we have to consider steady-state elongational (extensional) flow, and that strain rate is equal to elongational velocity gradient, $q$, in the direction of flow, $x$

$$\dot{\varepsilon} = d\varepsilon/dt = q = dV/dx \, . \quad (3)$$

Which of the above mechanisms is preferred depends on the molecular structure of the polymer.

Fluid flow and plastic deformation provide two extreme, ideal mechanisms of orientation. In real systems a mixture of deformation, and deformation rate effects can be expected.

Figures 1 and 2 illustrate developement of orientation in Nylon 6 fibers during melt spinning and cold drawing [2]. In Fig. 1, optical birefringence is plotted vs. average strain rate (characterized by velocity difference), in Fig. 2, birefringence is related to strain (draw ratio, spin-draw ratio, $V/V_0$).

The results indicate that orientation in cold drawing is practically independent of strain rate ($\Delta V$), but nearly linearily increases with strain, suggesting an ideal plastic, solid-state mechanism. On the other hand, orientation in melt-spinning, independent of strain but well correlated with strain rate (spinning velocity), points to a flow orientation mechanism.

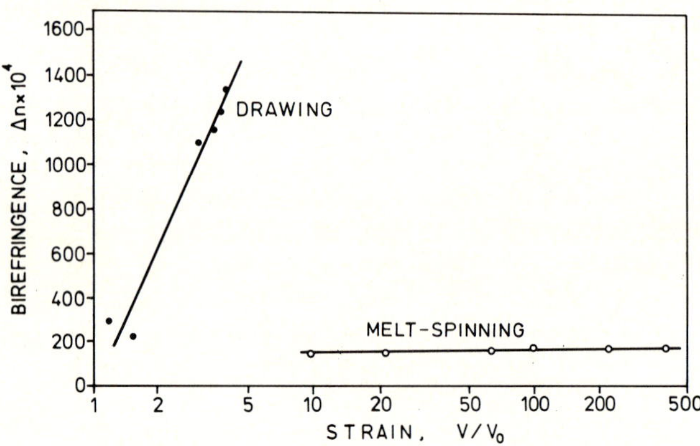


Fig. 1. Birefringence of melt-spun [17] and cold-drawn [18] Nylon 6 fibers vs. strain rate (velocity difference $\Delta V$). Strain (draw ratio, spin-draw ratio) constant

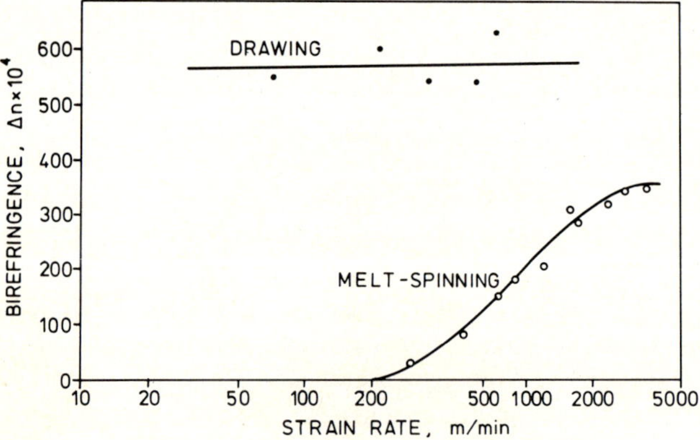


Fig. 2. Birefringence of melt-spun [17] and cold-drawn [18] Nylon 6 fibers vs. strain (draw ratio or spin-draw ratio, $V/V_0$). Spinning (drawing) velocity, constant

We will discuss the dynamic theory of orientation in more detail and compare the resulting conclusions with industrial processes.

## Molecular orientation in uniaxial deformation

The state of molecular orientation in a polymer system subjected to uniaxial deformation can be described by orientation distribution function $\Psi(\vartheta, t)$, where $\vartheta$ denotes orientation of a macromolecule (molecular segment) with respect to fibre axis. For condensed systems with intermolecular interactions, $\Psi(\vartheta, t)$ is determined by an integro-differential equation including intermolecular interactions in the mean-field approximation [3]

$$\frac{\partial \Psi}{\partial t} + \mathrm{div}_r \left[ \Psi \cdot \dot{\vartheta}_0 - D_r \cdot \left( \mathrm{grad}_r \Psi + \Psi \cdot \mathrm{grad}_r \cdot \frac{\int \Psi(\vartheta')\beta(\vartheta, \nabla')\, d\vartheta'}{kT} \right) \right] = 0 \,, \tag{4}$$

where $\dot{\vartheta}_0$ denotes convectional rotation velocity, and $D_r$ is rotational diffusion coefficient. $\vartheta$, in a general case denotes three Euler angles, and the differential operators $\mathrm{grad}_r$, and $\mathrm{div}_r$, are defined for the rotational (Riemannian) space [3, 4]. $d\vartheta'$ in the means-field integral should be understood as the appropriate volume element. In the simple uniaxial case, $\vartheta$ reduces to the angle between molecular and fiber axes, and the operators read

$$\mathrm{grad}_r F = (\partial F/\partial \vartheta) e_\vartheta \tag{5}$$

$$\mathrm{div}_r j = \frac{1}{\sin\vartheta} \frac{\partial}{\partial \vartheta} [\sin\vartheta\, j_\vartheta] \tag{6}$$

$$d\vartheta = 2\pi \cdot \sin\vartheta\, d\vartheta \,. \tag{7}$$

In elongational flow, rotation velocity $\dot{\vartheta}_0$ can be presented as a gradient of the flow potential, $Q(\vartheta)$,

$$\dot{\vartheta}_0 = \mathrm{grad}_r Q(\vartheta) = -\frac{3}{2} Bq \cdot \sin\vartheta \cos\vartheta \,, \tag{8}$$

where $B$ denotes shape factor, Equation (4) reduces to

$$\frac{\partial \Psi}{\partial t} - D_r \mathrm{div}_r \left[ \Psi \cdot \mathrm{grad}_r \cdot \left( \ln \Psi - Q/D_r + \frac{\int \Psi(\vartheta')\beta(\vartheta, \vartheta') \sin\vartheta'\, d\vartheta'}{kT} \right) \right] = 0 \,. \tag{9}$$

The degree of orientation, or axial orientation factor, $f_{or}$, is defined as a moment of the distribution function $\Psi(\vartheta, t)$

$$f_{or}(t) = \int P_2(\vartheta)\Psi(\vartheta,t)\sin\vartheta d\vartheta = \frac{3}{2}\langle\cos^2\vartheta\rangle - \frac{1}{2}, \tag{10}$$

where $P_2(\cos\vartheta)$ denoes second Legendre polynomial.

In our early analyses of the orientation problem [1], intermolecular interactions were neglected, and an exact solution of the linear orientation equation was obtained in the form of a series of exponential functions. The non-linear equation (9) dos not permit separation of variables.

A method of obtaining steady-state solutions of Eq. (9) has been proposed in ref. [4]. The "equilibrium" solution, $\Psi_{eq}(\vartheta)$, is controlled by two parameters:

— the ratio of strain rate to rotational diffusion coefficient, $q/D_r$;
— the intensity of intermolecular interactions, related to the function $\beta(\vartheta', \vartheta)$.

We will discuss two approximate solutions of the non-linear equation valid, respectively, for very large, and very small molecular mobility (diffusion coefficient).

For very large mobility, $D_r \to \infty$, orientation function in the mean-field functional can be replaced by equilibrium distribution, $\Psi_{eq}(\vartheta)$ and, after integration, reduced to a function of orientation $P(\vartheta)$

$$P(\vartheta) = \int \Psi_{eq}(\beta(\vartheta,\vartheta')\sin\vartheta' d\vartheta' . \tag{11}$$

With the mean-field integral replaced with the function $P(\vartheta)$ the variables $\vartheta$ and $t$ can be separated, and the approximate solution, similar to the solution of a linear equation, is obtained in the form

$$\Psi(\vartheta,t) \cong \sum_i \Psi_i(\vartheta)\exp(-\lambda_i D_r t) . \tag{12}$$

Sine $D_r$ is large, we will retain only the smallest positive eigenvalue, $\lambda_1$. Applying the initial condition

$$\Psi(\vartheta, t=0) = \Psi_0(\vartheta) , \tag{13}$$

we obtain

$$\Psi(\vartheta,t) \cong \Psi_{eq}(\vartheta; q/D_r) + [\Psi_0 - \Psi_{eq}]\exp[-\lambda_1 D_r t] . \tag{14}$$

The corresponding orientation factor

$$f_{or}(t; q/D_r) \cong f_{eq}(q/D_r) + (f_0 - f_{eq})\exp[-\lambda_1 D_r t] \tag{15}$$

is controlled, first of all, by the ratio $(q/D_r)$, rather than $q$ or $D_r$ taken separately. At $(D_r t) \to \infty$, orientation factor approaches its "equilibrium" value. Equation (15) and its asyptotic form, $f_{or} = f_{eq}(q/D_r)$, describe flow mechanism of orientation, active in systems of high molecular mobility. The theory and experimental techniques for studying flow orientation form an important branch of polymer science [6, 7].

In the other asymptotic case, $D_r \to 0$, Eq. (4) reduces to

$$\frac{\partial\Psi}{\partial t} + \mathrm{div}_r[\Psi \cdot \dot{\vartheta}_0] = 0 . \tag{16}$$

For uniaxial extension, Eq. (16) assumes the form

$$\frac{\partial\Psi}{\partial t} = -\frac{3}{2} Bq \frac{\partial}{\partial z}[\Psi \cdot z(1-z^2)] , \tag{17}$$

where $z = \cos\vartheta$. The solution reads

$$\Psi(z,t) = \Psi_{str} = \frac{a}{2[a-(a-1)z^2]^{3/2}} , \tag{18}$$

where

$$a(t) = \exp[3qBt] = \exp[3B\varepsilon(t)] = [R(t)]^{3B} \tag{19}$$

is a measure of time-dependent strain.

First solutions of the above problem were obtained 60 years ago by Kratky [8] and Oka [9]. Orientation is produced by affine rotation of structural elements in the deformation field, and determined by time-dependent strain, $\varepsilon = qt$ (or deformation ratio, $R$). Orientation factor $f_{or}$ is a sole function of strain. For very thin rods ($B = 1$) eqs. (18)—(19) yield:

$$f_{or}(t) = f_{or}[R(t)] = \frac{2a+1}{2(a-1)} - \frac{3a\,atn(a-1)^{1/2}]}{2(a-1)^{3/2}} = \frac{2R^3+1}{2(R^3-1)} - \frac{3R^3 atn[(R^3-1)^{1/2}]}{2(R^3-1)^{3/2}} . \tag{20}$$

Strain-controlled orientation factor as a function of draw ratio, $R$, is shown in Fig. 3.

A more general solution of Eq. (9), admitting small diffusional effects, will be obtained by perturbation around $\Psi_{str}$ with a small parameter $D_r/q$. The result can be presented in the form

$$\Psi(\vartheta,t) = \Psi_{str}[\vartheta,\varepsilon(t)] + (D_r/q)\Psi_1(\vartheta,t) + \ldots \quad (21)$$

The first correction function $\Psi_1$ is a solution of the equation

$$\frac{\partial \Psi_1}{\partial t} + \mathrm{div}_r[\Psi_1 \cdot \dot{\vartheta}_0] = F(\vartheta,t) = q \cdot \mathrm{div}_r \cdot \left[ \mathrm{grad}_r \Psi_{str} + \frac{\Psi_{str}\mathrm{grad}_r \int \Psi_{str}(\vartheta',t)\beta(\vartheta,\vartheta')\sin\vartheta' d\vartheta'}{kT} \right] . \quad (22)$$

Comparing the approximate solutions, we arrive at two ideal orientation mechanisms:

- strain orientation (Eqs. (16)—(22)) when molecular mobility $D_r$ is small, and orientation is a function of actual strain, independent (or slightly dependent) of strain rate mobility,
- flow (streaming) orientation (Eqs. (14)—(15)) at high mobility, where orientation is controlled by the ratio of strain rate to diffusion coefficient, $q/D_r$.

Experimental results shown in Figs. 1 and 2 demonstrate that orientation of Nylon 6 fibers in cold-drawing behaves in the way predicted for an ideal strain mechanism, while melt-spinning of the same polymer resembles equilibrium flow orientation. In other polymers, deviation from the ideal behavior, i.e., diffusional effects in drawing, and strain effects in spinning, may appear to be stronger [10—12].

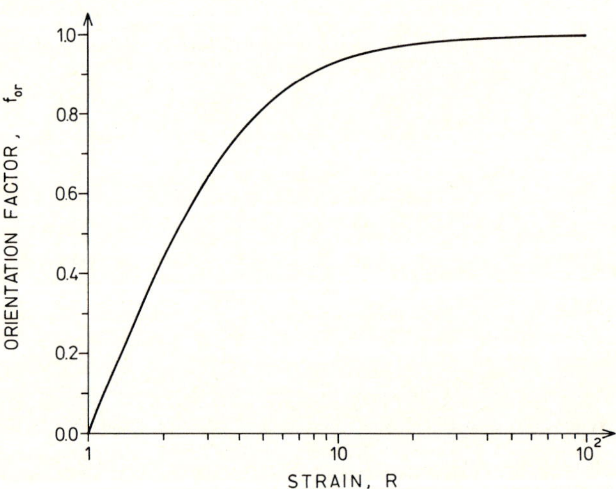


Fig. 3. Strain-controlled orientation factor for thin rods as a function of draw ratio, $R$ (cf. Eq. (20))

## Orientation and stress

Mechanical processing requires application of stress. In a fluid subjected to elongational flow, normal stress difference is proportional to strain rate, $q$

$$\Delta p = p_{11} - p_{22} = \eta_{el} \cdot q , \quad (23)$$

where $\eta_{el}$ is elongational viscosity. In an incompressible Newtonian fluid, $\eta_{el}$ is a constant equal to $3\eta_0$ ($\eta_0$ is Newtonian viscosity). In more general, non-Newtonian fluids, elongational viscosity is a function of strain rate $q$, and stress can be written in the form

$$\Delta p = \eta_{el}(q) \cdot q . \quad (24)$$

Equation (24) is also valid for steady-state elongational flow of viscoelastic fluids, effective elongational viscosity resulting from viscous and elastic effects. A similar equation can be written for plastic flow. Above the plasticity limit, $p_0$, stress is controlled by strain rate and "plastic viscosity", $\eta_{pl}$

$$\Delta p - p_0 = \eta_{pl}(q) \cdot q . \quad (25)$$

In all the above cases, stress is controlled by strain rate, $q$. A different situation occurs in elastic bodies where stress is controlled by strain. However, materials incapable of flow are never used for manufacturing of oriented fibers, and are beyond our interest.

In the strain-controlled mechanism, orientation depends on strain, while stress is controlled by strain rate. High degree of orientation can be produced without application of high stress, if the strain rate is low. This is exactly what happens when solidifed fibers (e.g., from UHMW polyethylene) are subjected to drawing in the plastic state.

On the other hand, in fluid-state processing both orientation and stress are controlled by strain rate. Steady-state orientation factor is a function of the ratio $q/D_r$

$$f_{eq} = a_1(q/D_r) + a_2(q/D_r)^2 + \ldots \quad (26)$$

Since rotational diffusion coefficient is inversely proportional to viscosity, also stress in Eqs. (23)—(24) can be expressed as a series

$$\Delta p = b_1(q/D_r) + b_2(q/D_r)^2 + \dots \tag{27}$$

which implies orientation-stress relation:

$$f_{eq} = C_{or}\Delta p[1 + c_2\Delta p + c_3\Delta p^2 + \dots] . \tag{28}$$

Such characteristics, typical for polymer fluids (melts, solutions, suspensions) do not exist in solid, plastic materials.

Unlike in the strain-controlled mechanism, flow orientation requires a definite stress level. The shape of the orientation-stress relation is a material property which plays an important role in fluid-state processing [10—12].

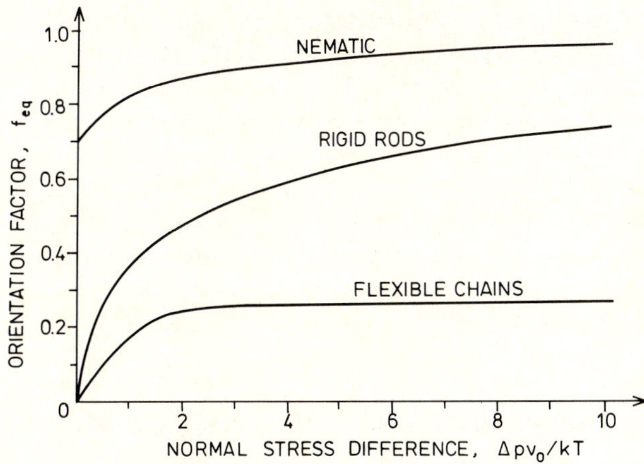


Fig. 4. Flow-controlled orientation-stress characteristics for various polymer fluids

## Orientation-stress characteristic and molecular rigidity

The uniqueness of orientation-stress relations has been discussed in our earlier papers [10, 11, 13, 15]. The functions $f_{or}(\Delta p)$ become unique material characteristics in steady-state elongational flow [11]. We have analyzed orientation-stress behavior for several special cases, including melts (or solutions) of flexible chain polymers [14], suspensions of rigid rods [13] and nematics [12]. Figure 4 presents typical results plotted vs. dimensionless normal stress, $(\Delta p v_0/kT)$.

The shape of the orientation-stress characteristics is crucial for fluid-state processing. All the characteristics are non-linear, and asymptotically approach ideal orientation ($f_{or} = 1$) at $\Delta p \rightarrow \infty$. The steeper the increase of orientation, the easier it is to produce the desired degree of orientation at a reasonably low stress level.

For flexible-chain polymers, orientation is a slowly increasing function of stress. Reaching high enough orientation may require application of a stress higher than tensile strength of the material. For flexible-chain materials, the shape of the $f_{or}(\Delta p)$ characteristic is practically invariant to molecular weight or chemical constitution of the polymer [10, 14]. The characteristics for non-interacting rigid rods are generally steeper, and the more so, the higher is their molecular volume and asymmetry (aspect) ratio [10, 13]. Orientation-stress characteristics for nematics start at a non-zero level determined by intermolecular interactions [12].

It is evident that fluid-state orientation can be efficient only when applied to large, asymmetric, rigid particles, preferably forming nematic structures. Table 1 presents rigidity characteristics for three polymers used for high-performance fibers: polyethylene, p-aramide (Kevlar), and p-phenyleno-benzo-thiazole (PBT).

In the theory of wormlike chains, effective rigidity can be characterized with the ratio of the persistence length, a (a material characteristic proportional to "bending modulus" of the macromolecule), divided by the total (contour) length of the extended molecule, $L$

$$x = a/L; \quad x \in (0, \infty) . \tag{29}$$

Macromolecules with $x < 0.001$ are commonly classified as "flexible", those with $x > 0.1$ as "rigid".

Table 1. Rigidity characteristics for selected fiber-forming polymers

| Molecular characteristic | Regular polyethylene | UHMW polyethylene | Kevlar | PBT |
|---|---|---|---|---|
| Persistence length, AU | 5.8 | 5.8 | 200—600 | 1000—1200 |
| Molecular weight, M | $10^4$ | $10^6$ | 32,000 | 17,000 |
| Contour, length L, AU | 900 | $9 \cdot 10^4$ | 2000 | 800 |
| Rigidity parameter $x = a/L$ | $6.4 \cdot 10^{-9}$ | $6.4 \cdot 10^{-5}$ | 0.1—0.3 | 1.25—1.5 |

Considering the mechanisms of orientation discussed in the preceding section, Kevlar and PBT are likely candidates for fluid-state orientation. Their orientation-stress characteristics should be very steep, the more so since both polymers in concentrated solutions form nematic structures. In flexible-chain polyethylene (regular or ultra-high-molecular-weight) fluid state orientation would be impractical because it requires too high a stress (cf. the appropriate curve in Fig. 4).

## Fiber-forming materials and processing conditions

Consideration of the orientation mechanisms points to three groups of materials, and three associated processing conditions:

i) fluid-state processing of rigid-rod polymers consisting of wet-, or dry-jet spinning. Best results are obtained with liquid-crystalline (nematic) dopes (PBT, Kevlar). The characteristic features of this process include high orientation and low stress related to steep orientation-stress relations. Unsucessful attempts of obtaining highly oriented fibers by high-speed melt spinning of flexible-chain PET [16] confirm the importance of molecular rigidity.

ii) Solid-state processing of high-molecular-weight, flexible-chain polymers — primarily polyethylene, but also polyacrylonitrile, polyvinyl alcohol, and polyoxymethylene. The processing includes formation of relatively loose, unordered structures via gel spinning, followed by slow drawing to large draw ratios in the plastic state. Flat orientation-stress characteristics, do not play any significant role.

iii) Two-step processing (fluid-state spinning followed by solid-state drawing) of semi-rigid polymers of various kinds. Semi-rigid polymers are more tractable: they can be dissolved in less aggresive solvents and/or fused in lower temperatures than rigid-rod aramides and heterocycles. On the other hand, reduced rigidity eliminates nematic structures, makes orientation-stress characteristics flatter, and reduces effects of flow orientation. Therefore, processing of semi-rigid polymers involves only partial orientation in the fluid state, which has to be completed in the solid state. Two-step processing (spinning + draw-

Table 2. High-performance fibers and mechanisms of their formation

| | Rigidity parameter $x = a/L$ | Tenacity G/den | Modulus G/den | Elongation % | Melting temp. °C | Manufacturing process | Orientation mechanism |
|---|---|---|---|---|---|---|---|
| PBT | 1.25—1.5 | 19.3 | 2173 | 1.2 | >600 | wet spinning LC dope | flow orientation |
| p-aramide (KEVLAR) | 0.1—0.3 | 28 | 966 | 2 | 570 | wet spinning LC dope | flow orientation |
| m-aramide (NOMEX) | | | 120 | | 430 | wet spinning isotropic dope | flow orientation |
| copolyamide (TECHNORA) | | 22 | 570 | 4.4 | 500 | wet spinning isotropic dope, drawing | flow orientation, strain orientation |
| UHMW PE (DYNEEMA) | $6.4 \cdot 10^{-5}$ | 34 | 1368 | 2 | 130 | gel spinning deep drawing | strain orientation |
| copolyester (VECTRA, HM) | | 16 | 483 | 3.4 | <350 | melt spinning thermotropic LC, drawing | flow orientation, strain orientation |

ing) was the earliest way of obtaining strong fibers from cellulose (Fortisan), aliphatic polyamides (Nylon 66) and polyesters (Terylene).

Table 2 presents examples of high-performance fibers made from flexible (PE), rigid (PBT, Kevlar) and semi-rigid polymers (Nomex, Technora, Vectra). It is evident that strongest fibers are manufactured either from rigid molecules wet-spun from nematic dopes (PBT, Kevlar), or from flexible UHMW polyethylene, gel-spun and drawn in the plastic state to high draw ratios. Processing of semi-rigid polymers consists of spinning from isotropic solutions (Technora, Nomex) or melts (Vectra), followed by drawing in the plastic state. Processing of flexible-chain polymers (polyolefines, PAN, PVA, polyoxymethylene) into highly oriented fibers, films or rods, always requires solid-state deformation. In addition to drawing of gel-spun UHMW PE fibers, also hydrostatic extrusion, die drawing, and tensile drawing applied to polyolefines and polyoxymethylene of medium molecular weight, result in high-tenacity and/or high-modulus materials.

## References


1. Ziabicki A (1970) J Soc Fiber Sci & Techn Japan 26:147
2. Ziabicki A (1979) In: Happey F (ed) Applied Fibre Science. vol III, Academic Press, London, pp 235—274
3. Ziabicki A (1991) Arch Mechanics 43:57
4. Ziabicki A (1990) Arch Mechanics 42:703
5. Ziabicki A, Jarecki L (1978) Colloid Polym Sci 256:332
6. Tsvetkov VN (1964) In: Ke B (ed) Newer Methods of Polymer Characterization, Interscience
7. Yamakawa H (1971) Modern Theory of Polymer Solutions, Harper & Row, New York
8. Kratky O (1933) Kolloid Z 64:213
9. Oka S (1939) Kolloid Z 86:243
10. Ziabicki A (1988) Structure development in processing polymers from rigid-rod vs. flexible-chain polymers, In: Lewin M (ed) Polymers for Advanced Structures, VCH Publishers, New York
11. Ziabicki A (1992) Orientation-Stress Relations of Polymer Fluids, Networks and Liquid Crystals Subjected to Uniaxial Deformation, In: Kuchanov A, Dusek K (eds) Polymer Networks '91, VSP, Zeist, pp 147—157
12. Ziabicki A (1991) Proceedings of the Fiber Producer Conference, Greenville, SC
13. Ziabicki A, Jarecki L (1988) Rheol Acta 26:83 (suppl)
14. Ziabicki A, Jarecki L (1986) Colloid Polym Sci 264:343
15. Ziabicki A (1990) Stress-Orientation Relations in Polymer Fluids, In: Oliver DR (ed) Third European Rheology Conference, Elsevier — Appl Sci Publ, London—New York, pp 534—536
16. Ziabicki A, Kawai H (1985) (eds) High-speed Fiber Spinning, Interscience, New York
17. Ziabicki A, Kedzierska K (1962) J Appl Polymer Sci 6:111
18. Arakawa S (1969) In: Formation of Fibres and Development of Their Structure, vol 1, Kagaku Dojin, Tokyo-Kyoto, p 287


Author's address:

Prof. Dr. hab. A. Ziabicki
Polish Academy of Sciences
Institute of Fundamental Technological Research
21 Swietokrzyska St.
PL-00-049 Warsaw, Poland

# Orientation and chain relaxation of amorphous polymers and compatible polymer blends

B. Jasse[1]), J. F. Tassin[2]), and L. Monnerie[1])

[1]) Laboratoire de Physicochimie Structurale et Macromoléculaire, associé au CNRS
Ecole Supérieure de Physique et de Chimie Industrielles de Paris, Paris, France
[2]) Laboratoire de Physicochimie Macromoléculaire, Université du Maine, Le Mans, France

*Abstract:* The technique of infrared dichroism was used to characterize segmental orientation developed above Tg in uniaxially stretched amorphous polymers and their compatible blends. The use of hydrogenated and deuteriated styrene block copolymers leads to a detailed analysis of the orientation relaxation along the chain and a quantitative comparison with the Doi-Edwards model has been performed. — By comparing the orientation relaxation of polystyrene and poly(methyl methacrylate), it has been shown that the friction coefficient is the suitable rescaling quantity. Under conditions where the friction coefficients are the same, both polymers behave identically. — Several compatible blends involving either polystyrene or poly(methyl methacrylate) have been investigated. In all the systems, the two polymers orient differently and the orientation depends on the blend composition. The change of the molecular weight between entanglements in the blend relatively to the pure components cannot account for the results. On the contrary, the increase in the mean friction coefficient in the blend, resulting from the interactions between the polymers leading to compatibility, appears as a major factor.

*Key words:* Orientation — infrared dichroism — polymer blends — polystyrene — relaxation

## Introduction

Polymer processing usually leads to molecular orientation of the macromolecular chains which, consequently, strongly influences the mechanical properties. Thus, a great deal of interest exists in the measurements of orientation and relaxation processes in order to correlate the processing conditions with the properties of the fabricated sample.

Numerous methods have been proposed to measure the orientation: x-ray diffraction, birefringence, sonic modulus, polarized fluorescence, broad line N.M.R., u.v. and infrared dichroism, and polarized Raman spectroscopy [1, 2]. These methods give information on the overall orientation or are specific to crystalline or amorphous regions. Vibrational spectroscopy is particularly attractive since it allows the possibility of obtaining information at the molecular level on the orientation of both crystalline and amorphous regions using specific vibrational modes associated with these different structures. Although the use of vibrational spectroscopy to measure orientation in polymers was reviewed [3], it is worthwhile to sum up briefly the information available from infrared dichroism measurements.

In the present paper, this technique will be used for describing the orientation and chain relaxation of amorphous homopolymers or miscible polymer blends stretched above their glass-rubber transition temperature.

## Infra-red dichroism

Absorption in the infrared region of the spectrum deals with vibrational motions of the various atoms of a molecule. These motions can be treated by performing a classical analysis of a vibrating system, which leads to a set of normal vibration frequencies associated with the normal modes of oscillation of the molecule. Owing to the change in the dipole moment of the molecule during the infrared active normal vibration, each mode will have a transition moment, $\vec{M}$, with a definite orientation in the molecule.

The orientation of a single unit of a polymer chain can be described by the three Eulerian angles, $\theta$, $\varphi$ and $\psi$ which define the three rotations required to bring into coincidence a set of cartesian axes in the oriented polymer. The orientation can be described by an orientation distribution function $f(\theta, \varphi, \psi)$. In the case of uniaxially-oriented systems, the orientation of structural units is assumed to be random with respect to $\varphi$ and $\psi$, and the orientation distribution function is expressed as:

$$f(\theta) = \sum_{n=0}^{\infty} (n + 1/2) \langle Pn(\cos\theta)\rangle Pn(\cos\theta) , \qquad (1)$$

where $Pn(\cos\theta)$ are the Legendre polynomials:

$$\begin{aligned} P_2(\cos\theta) &= (3 \cos^2\theta - 1)/2 \\ P_4(\cos\theta) &= (35 \cos^4\theta - 30 \cos^2\theta + 3)/8 , \end{aligned} \qquad (2)$$

etc.

For any absorption band, the dichroic ratio $R = A_{/\!/}/A_{\perp}$ ($A_{/\!/}$ and $A_{\perp}$ being the measured absorbance for electric vector parallel and perpendicular, respectively, to the stretching direction), is related to the second moment of the orientation function $\langle P_2(\cos\theta)\rangle$ by:

$$\langle P_2(\cos\theta)\rangle = \frac{(R - 1)(R_0 + 2)}{(R + 2)(R_0 - 1)} \qquad (3)$$

with $R_0 = 2 \cot^2 a$, $a$ being the angle between the dipole moment vector of the vibration and the chain axis, and $\theta$ is the angle between the chain axis and the draw direction. The angle $a$ can usually be obtained from theoretical considerations and dichroic ratio measurements allow the calculation of $\langle P_2(\cos\theta)\rangle$. Alternatively, if the orientation function value is known using a well-defined absorption band, the orientation of the dipole moment vector relative to the chain axis can be determined for any other absorption band.

Dichroism measurements are easily performed either on dispersive instruments or more conveniently, on a Fourier-transform apparatus. Measurements can be carried out either on stretched samples or during stretching. The main practical problem in measurements arises from the requirement of band absorbance lower than ca. 0.7 absorbance units in order to permit use of the Beer-Lambert's law. This means that one must obtain sufficiently thin films. Depending on the extinction coefficient of the considered band the required thickness can range from 1 to 200 μm. From this point of view, polymers with strong absorption bands (e.g. polycarbonate) are difficult to study.

The main difficulty in using infrared dichroism to study orientation is the necessity to find infrared absorption bands of the polymer which are sufficiently well assigned to normal vibrations of specified atomic groups. Such an assignment can be achieved by making a normal-coordinate analysis and experimentally by looking at deuteration effects and dichroic behavior. Furthermore, it is necessary that these well defined vibrational bands do not overlap with other bands resulting from another normal mode, an harmonic or a combination of other modes.

In vinyl polymers, on account of the different conformational structures existing in the macromolecular chain, the only univocal chain axis is the line joining two adjacent $CH_2$ groups. In polyphenylene oxide (PPO), the longest chain axis which gives identical values of $P_2$ using different absorption bands, in the experimental conditions used in this study is represented in Fig. 1b [4]. In any case, a linear relationship holds between birefringence and $P_2$.

On the other hand, in polymers such as polyethylene terephthalate (PET), it is impossible to define a univocal chain axis in the amorphous phase on account of the different conformational structures of the aliphatic part of the chain. In such a case infrared dichroism will lead to the orientation value of particular parts of the molecule.

Infrared dichroism studies do not require any labeling, however, consequently the derived orientation refers to an average over all the identical groups in a given polymer chain and over all the identical polymer chains present in the sample.

In order to get more detailed information on the orientation behaviour of the various parts of a polymer chain, deuterium labeling can be used to get block copolymers with hydrogenated and deuteriated blocks.

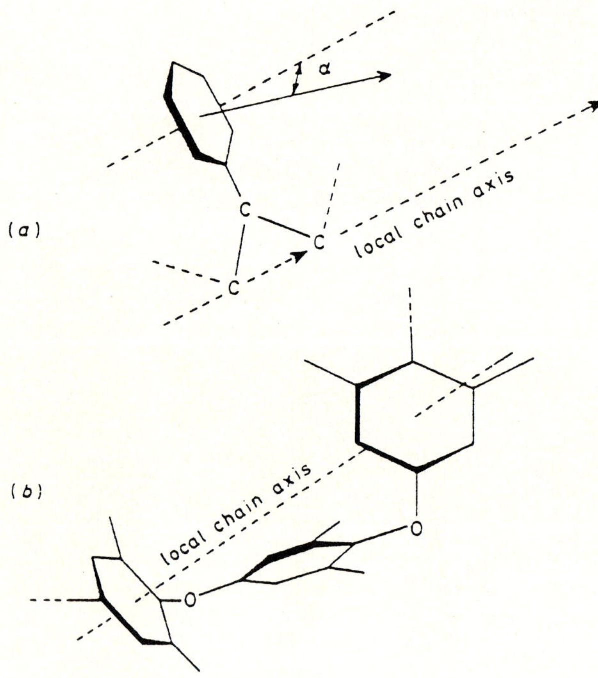


Fig. 1. Local chain axis a) in polystyrene and b) in poly(phenylene oxide)

In the case of blends, if specific bands which do not overlap can be found for each polymer, infrared dichroism is a good tool to study the orientation of each component in the blend.

## Methods

Thin films suitable for infrared measurements were obtained by casting solutions on a glass plate. Subsequent annealing was done in vacuum above the glass transition temperature in order to remove any trace of solvent and internal stress. The samples' dimensions for the stretching experiment were 40 mm × 17 mm × 0.05 mm. Unixaxial symmetry after stretching was confirmed by birefingence measurements. Samples for mechanical measurements were compression-moulded, as disks (diameter: 25 mm; thickness: 1.7 mm).

The glass transition temperatures of the samples were obtained using a Dupont 1090 differential scanning calorimeter. Heating rate: 20 °C · min$^{-1}$. Oriented samples from thin films were obtained on an apparatus developed in our laboratory [5], i.e., a stretching machine operating at constant strain rate and a special oven to obtain a very good temperature stability over the whole sample (homogeneity is ≈0.2 °C). The polarized spectra were recorded using either a Nicolet 7199 or a Nicolet 205 Fourier transform infrared spectrometer. The polarization of the infrared beam was obtained by the use of a gold wire grid polarizer. Dynamic shear experiments were obtained on a Rheometrics RMS 605 equipped with a plate-plate assembly.

## Homopolymers

In this section we first describe the results obtained on polystyrene, PS, looking at the effect of the molecular weight and the relaxation behaviour of the different parts of a chain (ends, central part). Results on poly(methylmethacrylate), PMMA, are presented and compared with those obtained on PS.

### *Polystyrene*

Different studies on PS orientation and relaxation have been performed using the 1028 and 906 cm$^{-1}$ absorption bands corresponding to the normal modes shown in Fig. 2.

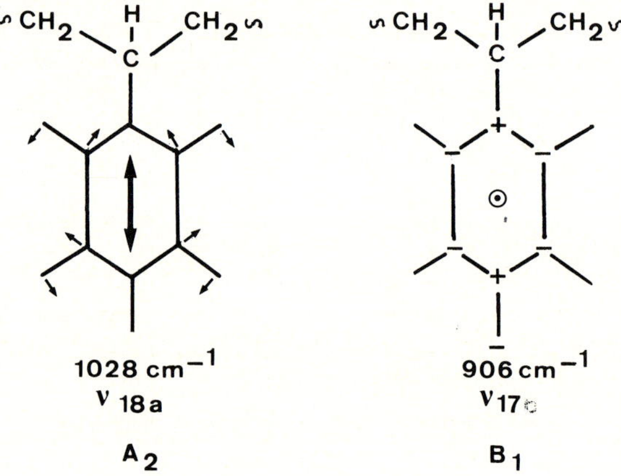


Fig. 2. Examples of normal vibration modes of polystyrene

When conventional PS is stretched at constant strain rate [6] a linear relationship holds between the orientation function $\langle P_2(\cos\theta)\rangle$ and draw ratio $\lambda = l/l_0$, where $l_0$ and $l$ are the lengths of the sample before and after stretching, respectively. Later, we use for characterizing the orientation, either the slope $d\langle P_2(\cos\theta)\rangle/d\lambda$ or the value of the orientation function $\langle P_2(\cos\theta)\rangle$ at a given draw ratio $\lambda$.

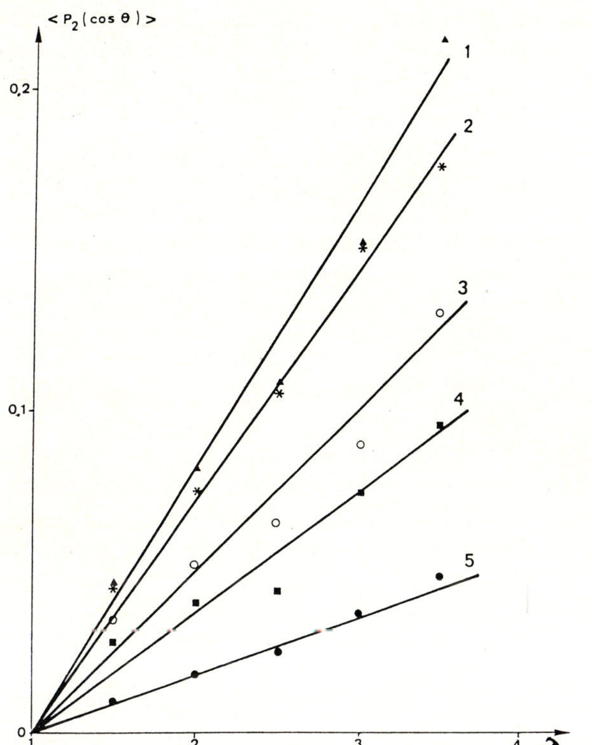


Fig. 3. Orientation functions of polystyrene as a function of draw ratio. Strain rate $\dot{\varepsilon}$ = 0.115 $s^{-1}$. Temperature of stretching: 1) 110°C; 2) 113°C; 3) 116.5°C; 4) 122°C; 5) 128.5°C

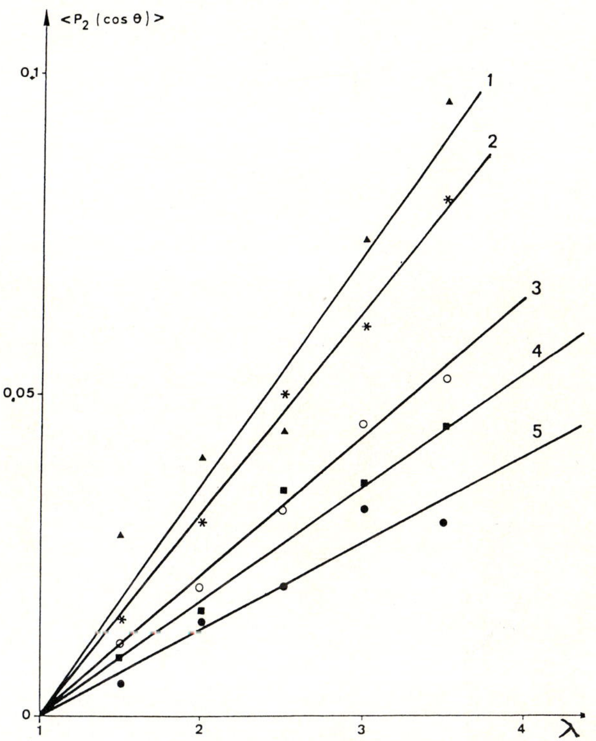


Fig. 4. Orientation function of polystyrene as a function of draw ratio. Temperature of stretching $T$ = 122°C. Strain rate: 1) 0.115 $s^{-1}$; 2) 0.086 $s^{-1}$; 3) 0.059 $s^{-1}$; 4) 0.026 $s^{-1}$; 5) 0.008 $s^{-1}$

For stretchings performed at a constant strain rate, $\dot{\varepsilon}$, the orientation decreases when the temperature increases, as shown in Fig. 3. In a similar way, orientation decreases when strain rate decreases for stretching at constant temperature, (see Fig. 4). These results suggest that relaxation processes are involved and it is tempting to check whether time-temperature superposition could be applied. Indeed, the orientation, measured at a given draw ratio, for a strain rate $\dot{\varepsilon}_1$ and a temperature $T_1$ is identical, within experimental uncertainty, with the orientation obtained at a temperature $T_2$ and a strain rate $\dot{\varepsilon}_2$ such that $\dot{\varepsilon}_2 = \dot{\varepsilon}_1 / a_{T_2/T_1}$, where $a_{T_2/T_1}$, is the shift factor between temperatures $T_2$ and $T_1$. It is given by the WLF equation:

$$\log a_{T_2/T_1} = \frac{-A(T_2 - T_1)}{B + (T_2 - T_1)} \quad . \tag{4}$$

The coefficient ($A$ = 9.06, $B$ = 69.8, $T_{ref}$ = 120°C) obtained from viscoelastic measurements have been used to lead to the orientation relaxation master curve at a reference temperature $T_1$ of 120°C, shown in Fig. 5, for the orientation measured at $\lambda = 4$ [7].

The master curves for the set of molecular weights investigated [7] are given in Fig. 6, using the same shift factors for all the polymers, as usual. It is worth noting two important points:

i) In the short time domain, orientation is independent on molecular weight.
ii) At longer times, orientation increases with the length of the chains, showing the appearance of a plateau for the highest molecular weight.

In order to get more details about the relaxation mechanisms of the various parts of a chain, orientation and relaxation studies have been performed on three block copolymers containing a deuteriated polystyrene block either at the end of the chain or in the middle of the chain [8]. The block copolymers and their molecular weight characteristics are enhanced in Fig. 7. Due to band overlaps of PSH (hydrogenated) and PSD (deuteriated), only the

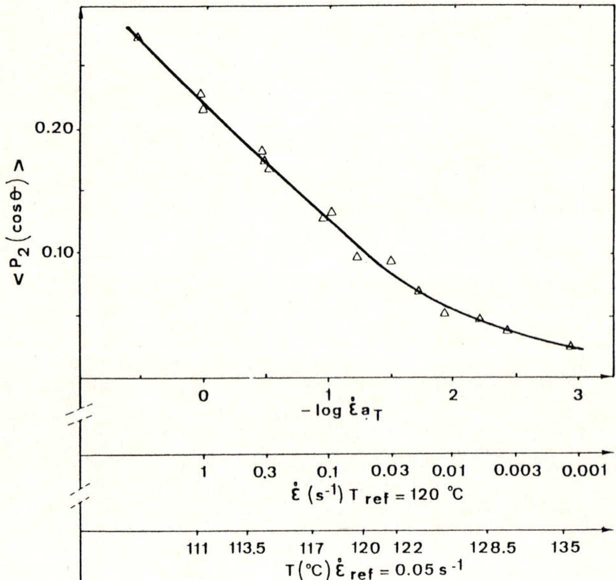


Fig. 5. Master curve of the relaxation of orientation for polystyrene at a reference temperature of 120 °C. The two axes below represent respectively the strain rate domain at a reference temperature of 120 °C and the temperature domain at an intermediate reference strain rate of 0.05 $s^{-1}$

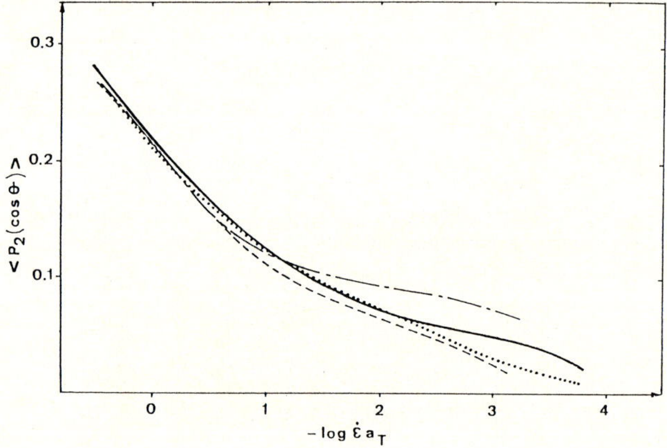


Fig. 6. Master curve of the relaxation of orientation for the narrow distribution polystyrene at a reference temperature of 120 °C (---) PS 100; (...) PS 180; (—) PS 500; (—.—) PS 900

group of bands lying between 2000 and 2300 $cm^{-1}$ can be used for PSD, in particular, the band at 2195 $cm^{-1}$ which corresponds to asymmetric stretching vibrations of $CD_2$ groups and that at 2273 $cm^{-1}$ which concerns stretching vibrations of the

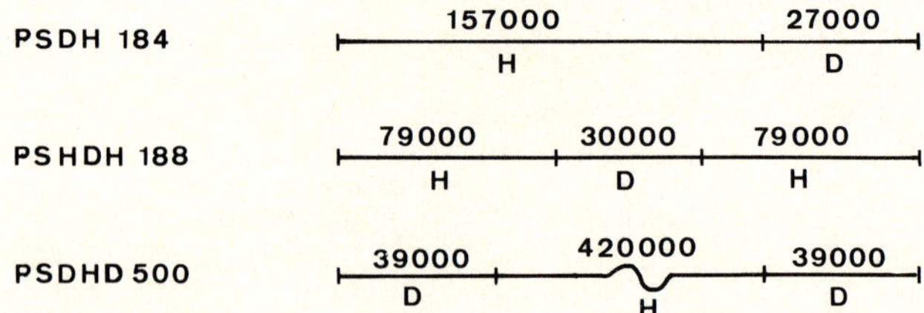


Fig. 7. Schemed structures and molecular weight characteristics of the H and D block polystyrene

aromatic C-D groups. For PSH, the band at 906 $cm^{-1}$ is convenient as it does not overlap with PSD bands.

The comparison between the relaxation of the orientation of the deuteriated end part of PSDH 184 and that of the overall chain is shown in Fig. 8. The orientation of the deuteriated end block appears to be always lower than the average orientation but the relative splitting between the two curves increases at long time, indicating a faster relaxation of chain ends. The smaller orientation of the end block at short times can be explained by the fact that no topological constraints are acting at the end of the chain to orient it.

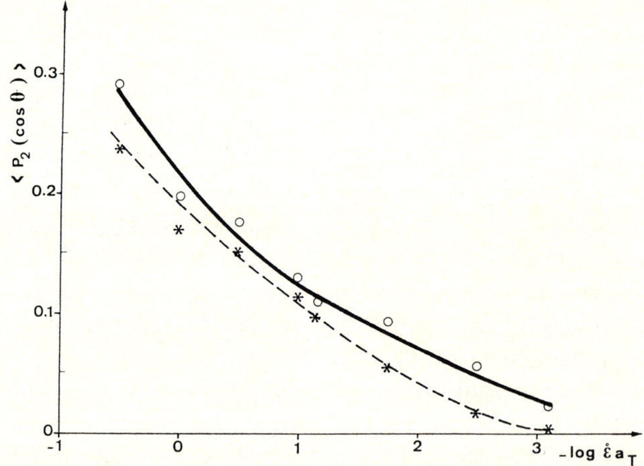


Fig. 8. Comparison between the master curves of the relaxation of average orientation (o, full line) and that of the deuteriated block (*, dashed line) for PSDH 184 at a reference temperature of 120 °C

The non uniformity of the relaxation along the chain appears even more clearly when the orientations of an end part and a central part are compared, as shown in Fig. 9, where the behavior of the deuteriated end block of PSDH 184 is compared to

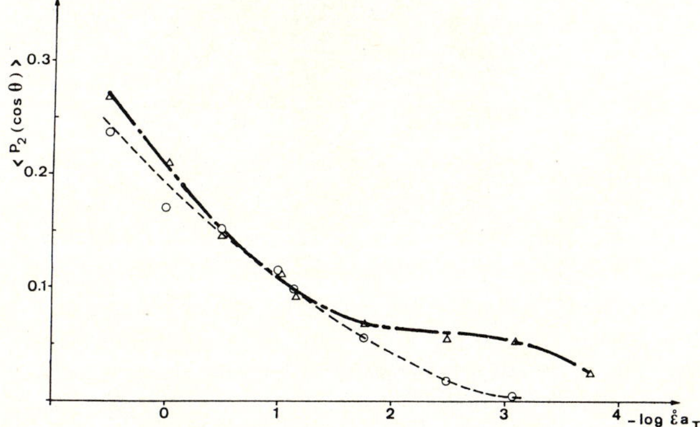


Fig. 9. Comparison between the relaxation of the deuteriated end block of PSDH 184 (*, dashed line) and that of the deuteriated central block of PSHDH 188 (△ alternate line)

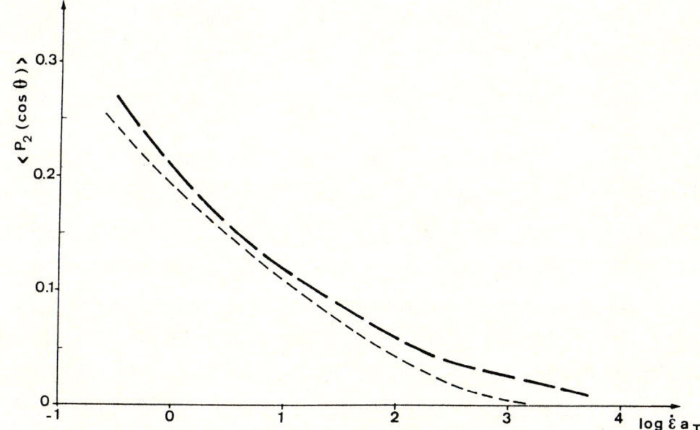


Fig. 10. Comparison between the relaxation of the deuteriated end block of PSDH 184 (dashed line) and that of the protonated block of PSHDH 188 (long dashed line)

that of the central deuteriated block of PS HDH 188. At short times (for —log $\dot{\varepsilon}a_T < 1$) the relaxation of orientation is independent on the location of the segments along the chain, whereas at longer times, the end block relaxes much more rapidly than the central part of the chain, the orientation of which goes through a plateau for $2 <$ —log $\dot{\varepsilon}a_T < 3$.

The influence of the length of the end part of the chain on its relaxation is depicted in Fig. 10, where the relaxation of the end parts of $M_w = 27\,000$ of PSDH 184 and of $M_w = 78\,000$ of PSHDH 188 are compared. The longer the end part, the slower the relaxation (at long times).

The relaxation at long times of the end part also depends on the overall chain molecular weight, as illustrated in Fig. 11, where the behavior of the deuteriated end blocks of the PSDH 184 and PSDHD 500 is plotted.

These results can be accounted for by considering the chain relaxation model developed by Doi and Edwards [9, 10]. In this model, the chain is confined inside a tube resulting from the constraints exerced by the surrounding chains. After stretching, the tube is affinely deformed and the relaxation of the chain towards the random coil conformation it gets in the isotropic state, occurs through three different processes at different time scales.

The first relaxation motion corresponds to a Rouse relaxation of a part of chain between two entanglements. This process is essentially local and its relaxation time ($\tau_A$) is independent on the molecular weight of the chain and on the location of the

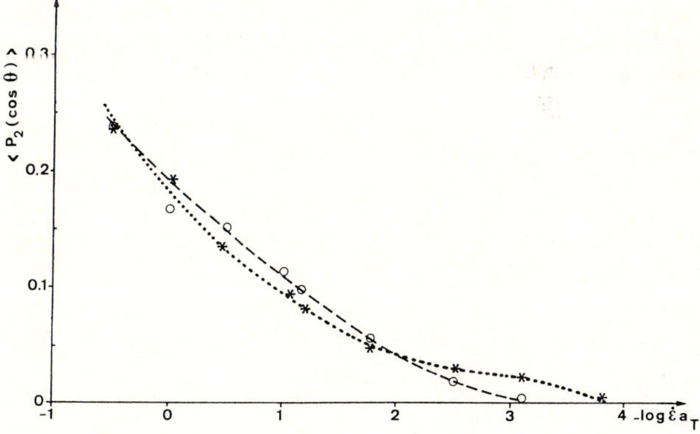


Fig. 11. Comparison between the relaxation of deuteriated end blocks: (*, dotted line); (○, full line) for PSDH 184

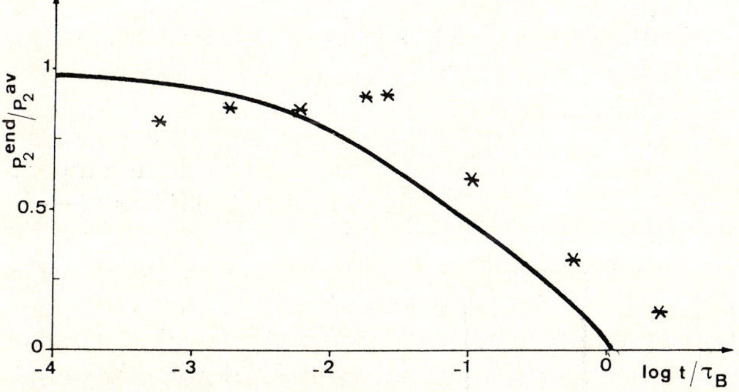


Fig. 12. Comparison between the experimental relative relaxation of the deuteriated end block of PSDH 184 (*) and the prediction of the retraction, chain length fluctuation, and reptation processes

segment along the chain. This is precisely what is observed in the short time range of Figs. 6, 9, 10 and 12.

The second relaxation process is a retraction of the deformed chain inside the deformed tube; its characteristic time ($\tau_B$) scales as the square of the molecular weight. The retraction motion first takes place at the ends of the chain and diffuses toward the center. The relaxation is therefore no longer uniform along the chain, as observed in Fig. 9. It is worthnoting that at the end of the retraction process, the orientation is constant along the chain and thus independent on the position of the segment.

The last relaxation motion corresponds to the reptation of the chain out of its deformed tube to reach an equilibrium isotropic conformation. As a basic assumption of this process, chain ends are assumed to lose their orientation as they leave the original deformed tube. Reptation therefore induces faster relaxation of chain ends. The disengagement time $\tau_{dis}^{D,N_0}$ by the reptation process of a labeled end part of a chain of length $N_D$ on a chain of global polymerization index $N_0$ scales as:

$$\tau_{dis}^{D,N_0} \propto N_0 N_D^2 \, . \tag{5}$$

Several improvements have been made to the original Doi-Edwards model, such as tube relaxation [11], chain length fluctuations [12—14].

To make a quantitative comparison between the theoretical models and the experimental data, it is convenient to study the time evolution of ratios of orientation of blocks and average orientation. Such ratios clearly exhibit the differences in the relaxation and they avoid any rescaling parameter between experimental and calculated values of the orientation.

The Doi-Edwards theory as well as the various implements which have been proposed, including chain length fluctuations, have been checked [8]. The best fit is obtained by considering the retraction process, the chain length fluctuations and the beginning of the reptation process. Such a fit is shown in Figs. 12 and 13 for the relative relaxations of the deuteriated end block of PSDH 184 and PSDHD 500 respectively. In both cases, the experimental data at long times are larger than the theoretical prediction. These differences can be interpreted by considering that the oriented surrounding induces an additional orientation of the chain end. This effect, which can be directly observed, as described later on, is more pronounced with the highest molecular weight PS which maintains a higher orientation at long times.

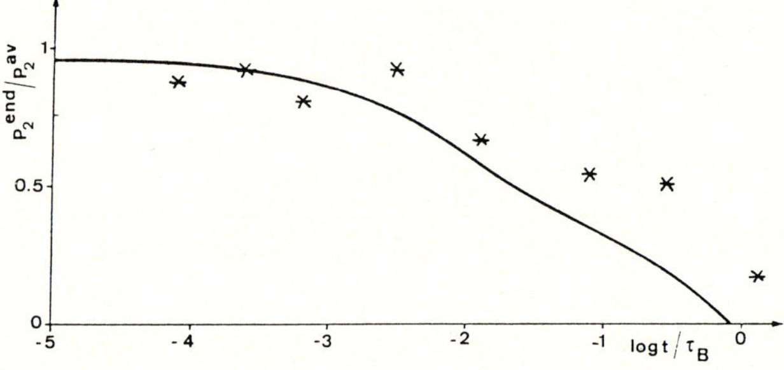


Fig. 13. Comparison between the experimental relative relaxation of the deuteriated end block of PSDHD 500 (*) and the predictions of the retraction, chain length fluctuation, and reptation processes

The scaling law predicted for the disengagement time (Eq. (5)) has been checked by comparing the long time relaxations of PSDH 184 and PSHDH 188. Experimentally, the splitting is of 0.9 logarithmic time units, corresponding to a ratio of the disengagement times of approximatively 8, in good agreement with the calculated value of 8.6. This accounts for the $N_D^2$ variation in Eq. (5). In a similar way, the proportionality with the polymerization index can be checked by comparing the deuteriated end blocks of PSDH 184 and PSDHD 500. The ratio between the disengagement times can be estimated as 5.5, whereas the scaling law yields 5.4.

As regards the orientation induced on a molecule or a polymer segment by the oriented surrounding, it can be directly evidenced by using a mixture of short deuteriated PS chains in long hydrogenated PS chains when the molecular weight of the short chains is lower than the molecular weight between entanglements (18 000 for PS), the short chains should remain unoriented by stretching for they cannot be entangled with the others. Such blends have been investigated by infrared dichroism [15] with short deuteriated chains of molecular weight 10 000 (PSD 10) at a concentration of 15% in matrices of long chains. With molecular weight 160 000 (PS 160) and 1 000 000 (PS 1000). The relaxation of the orientation functions measured at $\lambda = 4$ are plotted in Figs. 14 and 15. As it can be seen, the short chains are significantly oriented and in the long time range a measurable orientation remains in PS 1000. Such a behavior reveals the existence of a segmental orientational coupling with the oriented segments of the surrounding chains. A

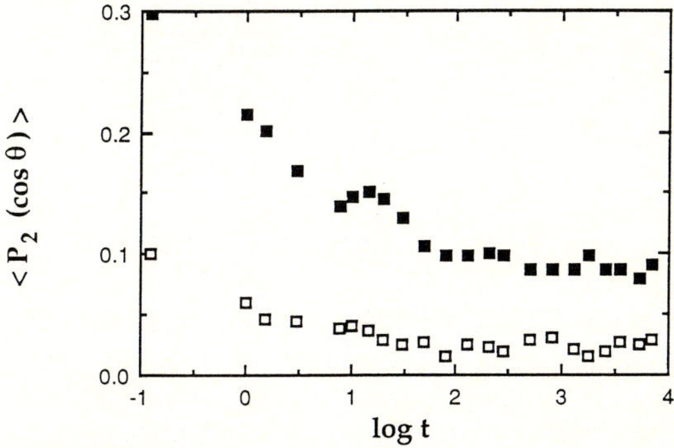


Fig. 14. Relaxation of the orientation of PS 1000 chains (■) and PSD 10 chains (△) in the blend PS 1000-PSD 10

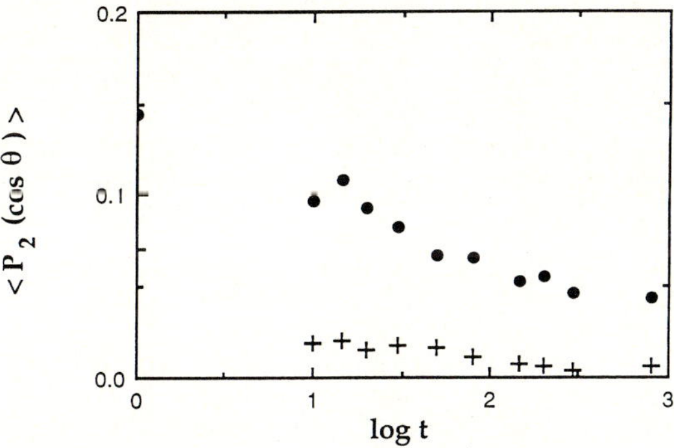


Fig. 15. Relaxation of the orientation of PS 160 chains (●) and PSD 10 chains (+) in the blend PS 160-PSD 10

model of chain orientation relaxation including this coupling has been developed by Doi et al. [16]. Our experimental results lead to an orientational coupling coefficient of 0.26.

*Poly(methyl methacrylate)*

PMMA orientation and relaxation was studied [17] using the 1388 and 749 $cm^{-1}$ absorption bands which are assigned to the symmetrical bending vibration of the α-$CH_3$ methyl group and to a skeletal vibrational motion affected by the $CH_2$ rocking vibration, respectively.

The general behavior of orientation as a function of strain rate and temperature is similar to PS, and an orientation relaxation master curve can be obtained by using the coefficients of the WLF equation obtained from viscoelastic measurements.

Comparison of orientation relaxation master curves of PMMA and PS [18] at reference tempera-

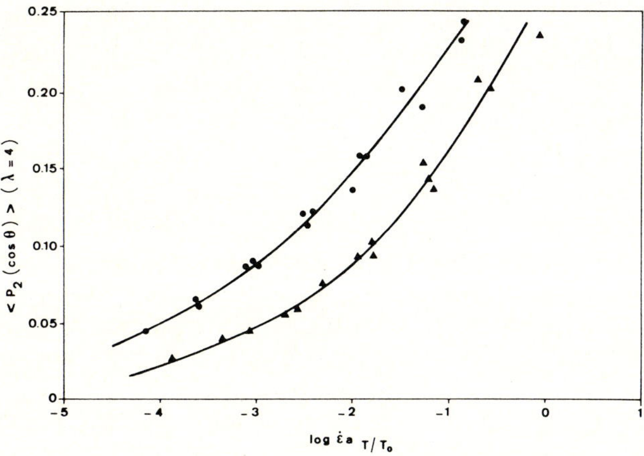


Fig. 16. Master curve of orientation relaxation of PMMA (●) and PS (▲) at a reference temperature $T_{ref}$ = Tg + 10°C

tures such that $T_{ref}$ = Tg + constant, shows that in these conditions, PMMA is more oriented than PS, as illustrated in Fig. 16.

A similar difference in orientation is maintained when the comparison is performed at the same value of the fractional free volume.

On the contrary, when the comparison is made at the same value of the friction coefficient, $\xi$, determined from viscosity [19] or viscoelastic measurements [20] and calculated at the appropriate temperatures by using the WLF equation with the corresponding coefficients, the orientation functions $\langle P_2(\cos\theta)\rangle$ at a draw ratio $\lambda = 4$ as a function of $-\log \xi$, fall on the same curve for a definite strain rate, as shown in Fig. 17.

Thus, the friction coefficient $\xi$, which depends on local intra-chain and inter-chain forces between neighbouring segments in the polymer melt seems to be a valuable reference parameter.

A study of the influence of the configurational structure of PMMA on orientation [21] has shown that, at the same strain rate and $T$ = Tg + constant, orientation decreases as follows: isotactic > syndiotactic > conventional.

These results are illustrated in Fig. 18. Orientation relaxation of isotactic and conventional PMMA is different, as it can be seen in Fig. 19. It is interesting to note that, similarly to the case of PS and PMMA the orientation relaxation of isotactic and conventional PMMA is identical when compared at the same friction coefficient, as it can be seen in Fig. 20.

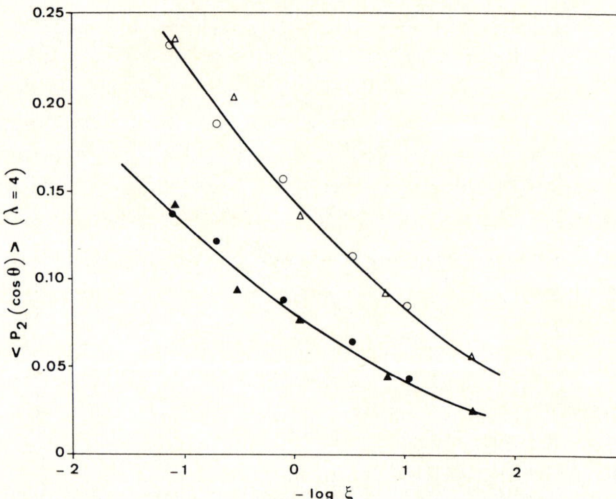


Fig. 17. Master curve of orientation relaxation of PMMA (○) and PS (△) at a same friction coefficient. Strain rate $\dot{\varepsilon}$ = 0.115 $s^{-1}$ (○, △); 0.08 $s^{-1}$ (●, ▲)

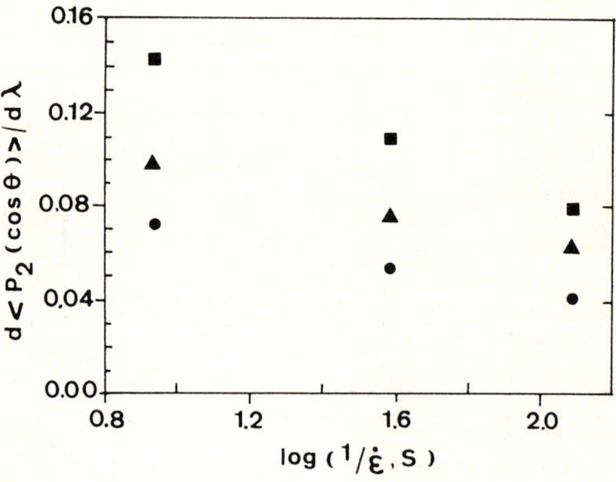


Fig. 18. Influence of strain rate on orientation of isotactic (■), syndiotactic (▲) and conventional (●) PMMA

At the same friction coefficient, we thus obtain the same relaxation curve for the various considered polymers. As a matter of fact, at high strain rate and a stretching temperature close to Tg, relaxation processes occurring during stretching are minimized and one should observe a difference in orientation due to the difference between the number of links, Ne, between entanglements.

The main factor to be considered is the number of Kuhn segments, Nk, between entanglements. This number can be calculated from the length $\langle L_k \rangle$ of a Kuhn segment, given by:

$$\langle L_k \rangle = \langle l_e \rangle C_\infty ,$$

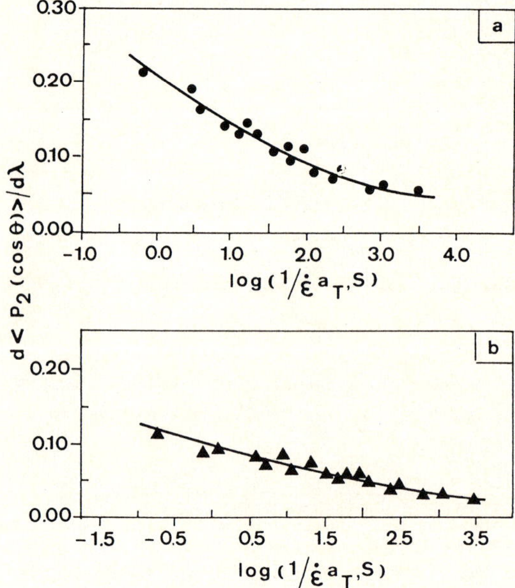


Fig. 19. Orientation relaxation master curves of isotactic (a) and conventional (b) PMMA at a reference temperature $T_{ref}$ = Tg + 11.5°C

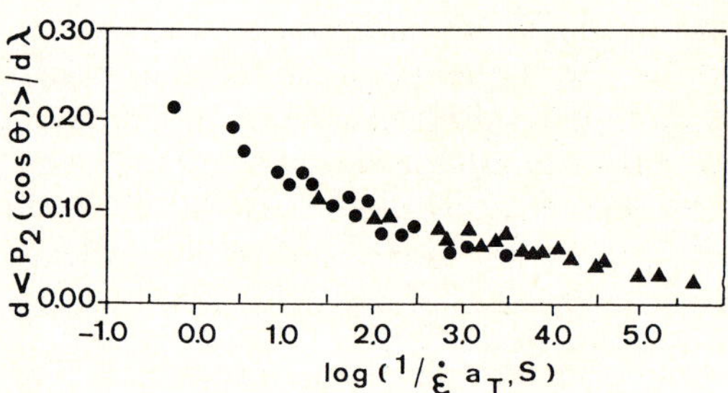


Fig. 20. Orientation relaxation of isotactic (●) and conventional (▲) PMMA at same friction coefficient

where $l_e$ is the length of a C-C link in the chain (1.53 Å) and $C_\infty$ the characteristic ratio. The number of Kuhn segments between entanglements $N_k$ is then given by:

$$N_k = \frac{L_e}{\langle L_k \rangle} ,$$

where $L_e$ is the chain length between entanglements.

$$L_e = N_e \langle l_e \rangle .$$

Using the data of Wu [22] for $C_\infty$ and Ne, the following values of Nk are obtained for PMMA and PS:

Table 1. Number of Kuhn segments, $N_k$

| | isotactic | conventional | syndiotactic |
|---|---|---|---|
| PMMA | 27 | 21 | 21 |
| PS | 40 | 33 | |

These results show that isotactic PMMA contains a higher number of Kuhn segments between entanglements than the syndiotactic polymer and should present a lower initial orientation. Similarly, isotactic PS should be less oriented than the conventional polymer.

All these observations clearly show that, in the initial orientation processes and in the first part of the orientation relaxation, the entanglements do not play any role, contrary to what is assumed in the Doi-Edwards model. Indeed, this model would predict a relaxation time $\tau_A$ four times shorter for conventional PMMA than for PS under conditions in which the friction coefficients are identical.

The relaxation processes which are involved in this first part of the chain relaxation are too local to be affected by the entanglements of the chains which will lead, at longer times, to the appearance of a plateau in the orientation relaxation (see Fig. 6). On the contrary, the relaxation times of these segmental motions are controlled by the friction coefficient of the polymer as it has been proven by spectroscopic techniques [23].

## Compatible blends

In a previous review on segmental orientation in multicomponent polymer systems, Wang and Cooper [24] concluded that compatibility at the molecular level usually leads to similar segmental orientation of the blend components. The studied systems were based on polymers allowing the establishment of intermolecular hydrogen bonding between different species. It is interesting to examine the behaviour of other compatible polymer blends which do not exhibit hydrogen bonding. In the different examples shown hereafter, the compositions are given as weight fractions.

### *Polystyrene-Poly(phenylene oxide)*

PS and PPO are well known to give compatible blends in the entire concentration range. The compatibility originates from intermolecular interactions characterized by a negative value of the Flory-Huggins interaction parameter $\chi$. In PS-PPO blends, similarly to the previous homopolymers, linear relationship holds between the orientation function $\langle P_2(\cos\theta)\rangle$ and the draw ratio $\lambda$. This seems to be a general feature in the different blends studied except for (PMMA-Poly(Trifluoro Ethylene)). The orientation of PS and PPO was measured [25] from the absorption band at 906 cm$^{-1}$ in PS and from those at 860 cm$^{-1}$ and 1195 cm$^{-1}$ in PPO which correspond to the out-of-plane $\nu_{11}$ mode of the benzene ring and to the asymmetric stretching vibration of the ether group, respectively. Figure 21 shows that the orientation of the two chains does not follow a similar path.

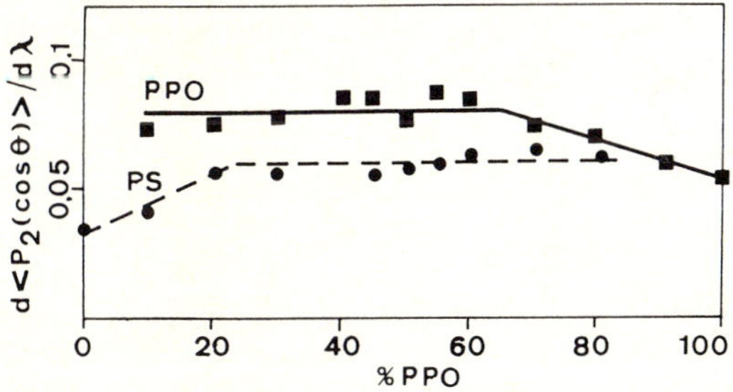


Fig. 21. Orientation as a function of the composition of PS-PPO blends. Stretching temperature $T$ = Tg + 11.5°C, strain rate $\dot{\varepsilon}$ = 0.026 s$^{-1}$

The introduction of a small amount of one component results in a linear increase in the orientation of the major component up to a limit concentration $C_1$. In the present case $C_1 \approx$ 20% of PPO for PS orientation and $C_1$ = 25% of PS for PPO orientation. Between these two values, orientation of both components is insensitive to the composition of the blend. It has also been shown than the $C_1$ concentration decreases with an increase in strain rate and a decrease in the stretching temperature [26]. Similarly, the viscoelastic plateau modulus increases with an increase in the PPO percentage in PS in the range 0—30% PPO [26].

The increase in orientation of PS when a small amount of PPO is added to this polymer could be explained by a possible decrease in entanglement molecular weight due to the presence of PPO chains [27]. However, the plateau observed between the two limit concentrations and the increase in

PPO orientation when a small amount of PS is added to this polymer, are not consistent with such an explanation. As a matter of fact, the entanglement molecular weight increases in the blends as compared with pure PPO. On the other hand, the increase in orientation observed for both polymers in the blends when compared with the pure components is in good agreement with an increase in friction coefficient due to intermolecular interactions between PS and PPO chains.

It is worth noting that PS and PPO have intrinsic birefringence, $\Delta n^0$ of opposite signs:

$\Delta n^0(\mathrm{PS}) = -0.10$

$\Delta n^0(\mathrm{PPO}) = +0.21.$

Consequently, depending on the composition of the blend, negative or positive birefringence develops with stretching. Thus, for a composition containing 27 weight % of PPO no birefringence appears by stretching [28].

*Polystyrene-Poly(vinyl methyl ether)*

Although PS-PVME blends are compatible over the whole concentration range ($\chi < 0$), the addition of a small amount of PVME involves a sharp decrease in the glass temperature of the blend but the stretching process requires a large difference between the experimental temperature and room temperature in order to freeze the orientation achieved. PVME composition in the blends is therefore limited up to 25%. The infrared spectrum of the blend shows that the absorption band of PS at 906 cm$^{-1}$ and that at 2820 cm$^{-1}$ of PVME, corresponding to the symmetric stretching of methoxy group, allow the determination of the orientation of each component.

In the investigated [29] concentration range a strong increases in PS orientation versus PVME concentration is observed, as shown in Fig. 22. As far as PVME is concerned, almost no orientation is achieved except at high draw ratios, as indicated by birefringence measurements. In these blends, the plateau modulus also increases with an increase in the PVME percentage in PS [30].

*Poly(methyl methacrylate)-Poly(ethylene oxide)*

PMMA-PEO blends belong to a category of polymer blends in which one of the components

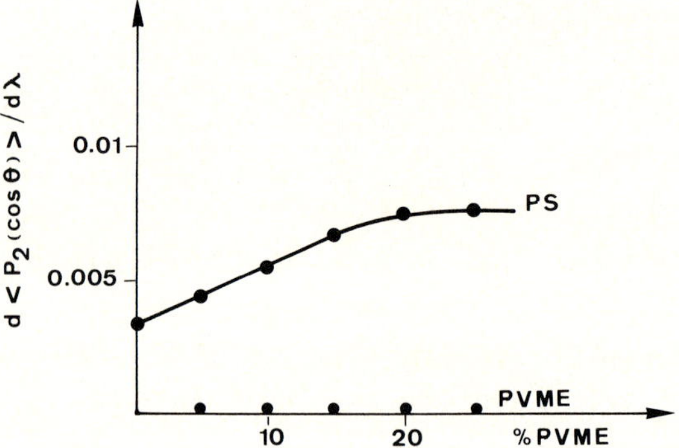


Fig. 22. Orientation as a function of the composition of the blends. Temperature of stretching $T$ = Tg + 115°C, strain rate $\dot{\varepsilon}$ = 0.115 s$^{-1}$

may crystallize-in this case PEO-. The negative value obtained for the Flory-Huggins $\chi$ interaction parameter shows that such blends are thermodynamically stable in the melt. In the solid state, the crystallization of PEO is influenced by PMMA. Up to a content of about 40 wt% of PMMA, the blend films are completely filled with PEO spherulites, the PMMA molecules being incorporated. When PMMA concentration increases, phase separation may occur, resulting in crystallites coexisting with two amorphous phases. No crystallization is observed in blends containing less than 20% of PEO. The glass transition temperature of this system is above the crystallization temperature and the blends are amorphous, transparent and exhibit only one glass transition temperature.

The infrared spectra of PMMA and PEO show that absorption bands of both polymers badly overlap and only PMMA orientation can be measured using the 749 cm$^{-1}$ absorption band, which is assigned to a skeletal vibrational motion affected by the $CH_3$ rocking vibration. The knowledge of PMMA orientation allows one to estimate PEO orientation from birefringence measurements.

The studies [31] performed on blends containing up to 20% of PEO show the concentration dependence of PMMA orientation reported in Fig. 23. PMMA orientation goes through a maximum for a concentration of 5% PEO, then decreases regularly, coinciding with pure PMMA orientation at a PEO concentration of 20% but at different temperature, because of $T_{rel}$. It seems that those peculiar behavior could be assigned to the initiation of phase separation.

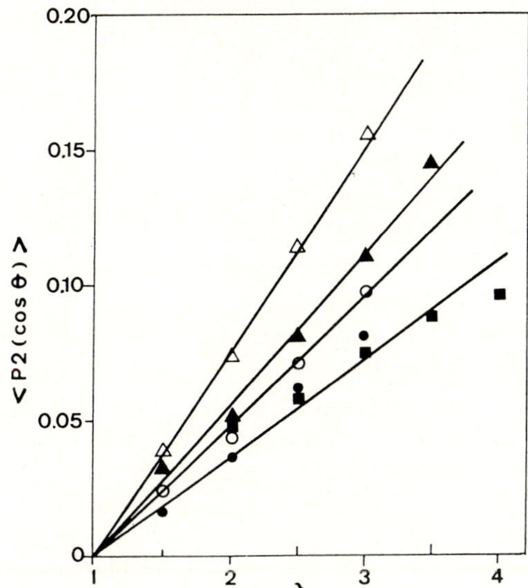


Fig. 23. PMMA orientation as a function of draw ratio in PMMA-PEO 20 000 blends: (■) pure PMMA; (△) 5% PEO; (▲) 10% PEO, (○) 15% PEO; (●) 20% PEO. Strain rate: 0.026 $s^{-1}$. Temperature of stretching $T$ = Tg + 21 °C

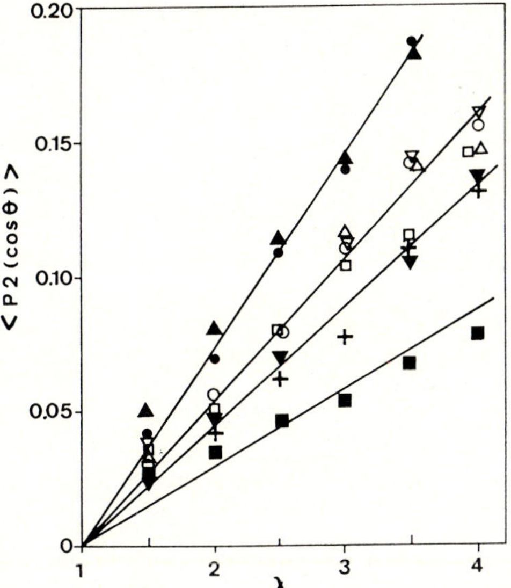


Fig. 24. Orientation of PMMA in PMMA (95%) — PEO (5%) blends as a function of PEO molecular weight: (+) 600; (▼) 1000; (□) 2000, (▽) 4000; (△) 20 000; (○) 50 000; (●) 200 000; (▲) 600 000; (■) pure PMMA. Strain rate: 0.026 $s^{-1}$. Temperature of stretching $T$ = Tg + 23 °C

As far as PEO is concerned, this polymer has a very short relaxation time which is not in the time range covered in the experimental conditions and no orientation can be observed.

Viscoelastic measurements also indicate that the mechanical relaxation is strongly hindered in the blend containing 5% of PEO. Then, when the PEO percentage increases, the hindrance in relaxation becomes less important.

The effect of PEO molecular weight on PMMA orientation is illustrated in Fig. 24. Three different orientation levels can be distinguished. The two higher molecular weights (200 000 and 600 000) lead to the larger increase in PMMA orientation and cannot be differentiated from one another. A second level of orientation is reached when PEO molecular weight decreases. However, no influence of molecular weight on PMMA orientation is detectable in the range 2000—50 000.

A third level of PMMA orientation is obtained with PEO molecular weight 600 and 1000, PMMA orientation being, however, higher than pure PMMA orientation.

PMMA orientation relaxation master curves obtained from infrared dichroism results are shown in Fig. 25 for pure PMMA and blends containing 5% and 10% of PEO 20 000. In the long relaxation time range, it is impossible to distinguish the two blends while, in the short relaxation time range, PMMA is more oriented in the 5% blend than in the 10% blend.

This increase in PMMA orientation in the blends cannot be related to an increase of chain entanglements. Indeed, Wu [27] has shown that in different compatible blends, including PMMA-PEO blends, the specific interactions tend to align the chain segments for the interaction, thus locally stiffening the segments, reducing their convolution and finally resulting in reduced entanglements between dissimilar chains.

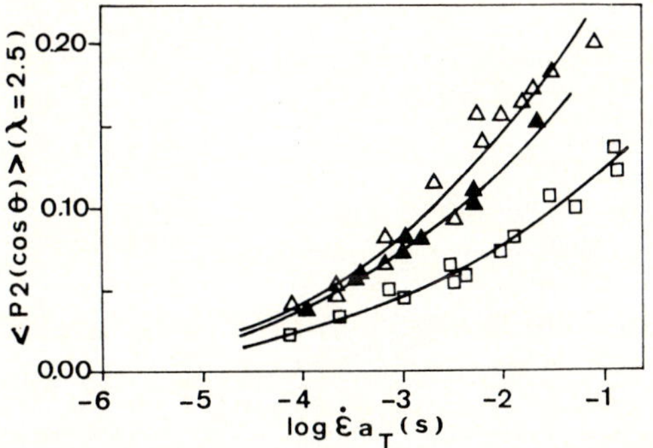


Fig. 25. Orientation relaxation master curve for PMMA-PEO 20 000 and a draw ratio $\lambda$ = 2.5. PEO content: (□) 0%; (△) 5%; (▲) 10%. Reference temperature $T_{ref}$ = Tg + 10 °C

*Poly(methyl methacrylate)-Poly(trifluoro ethylene)*

PMMA-$PF_3E$ blends are also blends in which one component can crystallize. Leonard et al. [32] have shown that intermolecular interactions in such blends appears to be stronger than in PMMA-$PVF_2$ blends ($\chi$ = —0.29 at 160 °C). Homogeneous blends are obtained for $PF_3E$ concentrations below 20% $PF_3E$.

For these blends, infrared spectra show that the absorption band of PMMA at 749 $cm^{-1}$ and that of $PF_3E$ at 540 $cm^{-1}$, tentitatively assigned to a $CF_2$ group vibration, can be used for measuring the orientation of each polymer.

Orientation studies have been performed in the concentration range leading to amorphous blends [33]. Values of the PMMA orientation function in the blends versus draw ratio $\lambda$, as a function of $PF_3E$ percentage are given in Fig. 26.

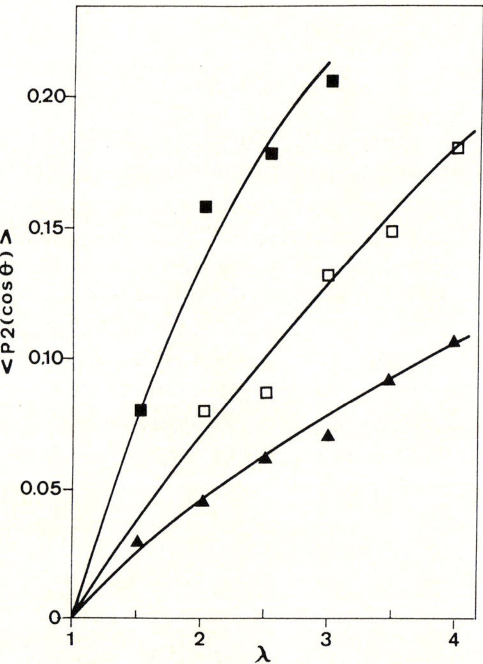


Fig. 26. Orientation of PMMA in PMMA-$PF_3E$ blends as a function of $PF_3E$ percentage. (▲) pure PMMA; (□) 5% $PF_3E$; (■) 10% $PF_3E$. Strain rate: 0.026 $s^{-1}$. Stretching temperature $T$ = Tg = 17 °C

One can notice that, in these particular blends, a linear relationship between $\langle P_2(\cos\theta)\rangle$ and $\lambda$ does not hold. Introduction of $PF_3E$ in PMMA leads to a progressive increase in the orientation of the PMMA chains. On the other hand, no significant difference is observable between the two blends for $PF_3E$ orientation, deduced from birefringence measurement.

The present results are in good agreement with the behavior previously observed in other compatible blends, that is, that orientation is concentration-dependent and increase when the corresponding component is present in a large amount.

*Poly(methyl methacrylate)-Copoly(styrene-co-acrylonitrile)*

In PMMA-SAN blends, the Flory-Huggins interaction parameter has a very small value ($\chi$ = —0.011) [34] and compatibility has mainly an entropic character. Blends are only compatible in a concentration range 9—28% of acrylonitrile in the copolymer. The present results concern a SAN containing 25% acrylonitrile.

Orientation studies of the pure SAN copolymer [21] based on the dichroism of the $CH_2$ stretching modes at 2930 and 2850 $cm^{-1}$ and of the in-plane ring mode at 1028 $cm^{-1}$ show that the two types of segments orient in the same way.

In the blend, the absorption band at 1028 $cm^{-1}$ was used for SAN and that at 1388 $cm^{-1}$ for PMMA.

The orientation of the two polymers in function of the blend composition [21] is shown in Fig. 27.

The first point to notice is that both polymers orient differently. As SAN percentage increases in the blends, PMMA orientation first remains almost constant then decreases regularly. Similarly, an increase in PMMA concentration in the blends has no effect on SAN orientation up to 30%, then SAN orientation increases regularly. It was also shown that orientation in both polymers decreases with increasing stretching temperature. Orientation relaxation master curves, established for the two polymers in different blends at a reference temperature $T_{ref}$ = Tg + 11 °C, are shown in Fig. 28. The general behavior of both components could be explained qualitatively by the change in molecular weight between entanglements, Me, observed by Wu [27].

Thus, an increase in the amount of SAN which induces an increase in the molecular weight between entanglements, also induces a fall off in the orientation of PMMA.

However, it is interesting to note that PMMA orientation remains almost constant up to 40%

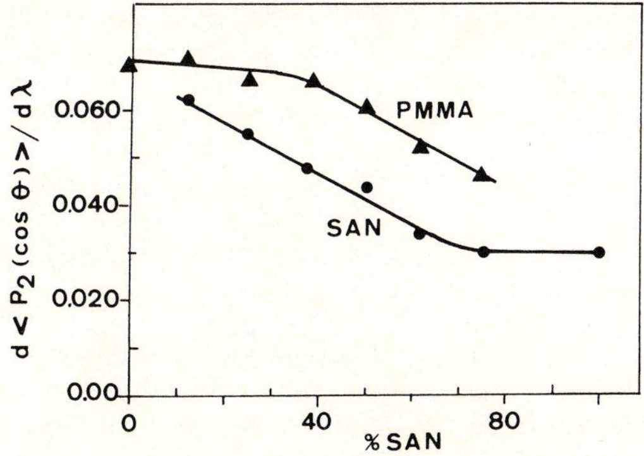


Fig. 27. Orientation of PMMA and SAN in PMMA-SAN blends as a function of SAN amount. Strain rate: 0.115 $s^{-1}$. Temperature of stretching $T$ = Tg + 14°C

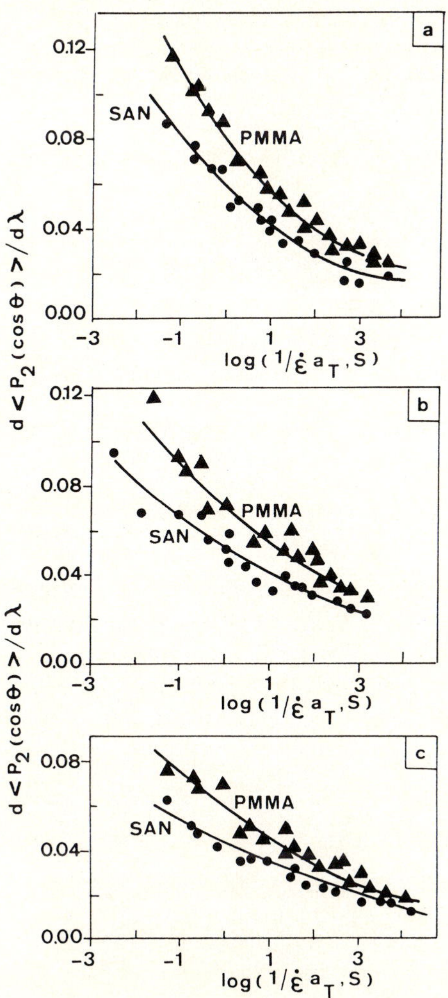


Fig. 28. Orientation relaxation of PMMA (▲) and SAN (●) in PMMA-SAN blends. Reference temperature $T_{ref}$ = Tg + 11°C. (a) 25% SAN; (b) 50% SAN; (c) 75% SAN

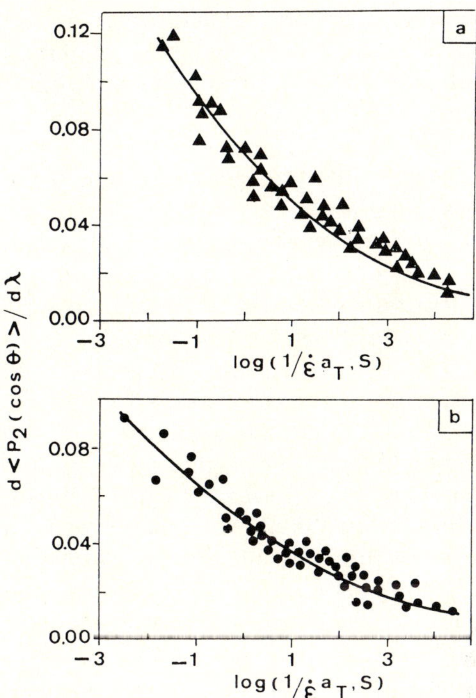


Fig. 29. General orientation relaxation master curves of PMMA (a) and SAN (b) in PMMA-SAN blends at same friction coefficient [Ln$\zeta$(dyne. s/cm) = 8.47]

SAN while Me increases from 9000 g · $mol^{-1}$ to 11 700 g · $mol^{-1}$, which makes doubtful, at least on a large scale, the influence of Me on orientation. On the other hand, when the results are compared at the same value of the mean friction coefficient, calculated from the values given in ref. [27], one obtains two master curves (Fig. 29) representative of the orientation relaxation behavior of PMMA and SAN in blends, whatever the concentration.

## Conclusions

Several general features of amorphous polymer orientation can be drawn from the previous results.

1) Studies performed on PS have shown that the short time range part of the relaxation is independent on the chain molecular weight. Furthermore, a detailed analysis of the relaxation mechanism shows that the relaxation does not occur uniformly along the chain sequence. It is possible to account for the faster relaxation of the ends relatively to the average orientation by considering the improved Doi-Edwards model, including chain retraction, chain length fluctuations and reptation.

2) Comparison of orientation relaxation master curves of PS and PMMA show a higher orientation of PMMA when the comparison is performed at reference temperatures such that $T_{ref}$ = Tg + constant.

On the contrary, under conditions in which the two polymers have the same friction coefficient, their orientation behaviors are identical. This proves that the convenient rescaling for relaxation is through the friction coefficient, in agreement with what has been already observed for local dynamics in polymer melts [23]. Such a result contradicts the assumption of Doi-Edwards that in the first relaxation process the relaxation time is controlled by the molecular weight between entanglements.

3) In compatible amorphous blends, different behavior are observed. When the blends possess a negative value of the Flory-Huggins interaction parameter $\chi$, both polymers orient differently. Orientation is concentration-dependent and increases linearly when the corresponding component is present in a large amount. This behavior cannot be taken into account by a change in molecular weight between entanglements. As a matter of fact, interactions greatly modify the molecular environment of polymer chains. Therefore, mobility is hindered and the friction coefficient of each species is increased. The increase in orientation observed in these blends when compared to the pure components is in good agreement with an increase in friction coefficient.

In blends where inter-chain interactions are almost inexistent, both components usually orient differently. The concentration dependence of orientation is not the same as in the previous case but the mean friction coefficient allows one to rescale the orientation relaxation of each component, independently of the blend composition.

## References


1. Samuels RJ (1974) Structured Polymer Properties, Wiley, New York
2. Ward IM (1975) Structure and Properties of Oriented Polymers, Applied Science Publ, London
3. Jasse B, Koenig JL (1979) J Macromol Sci, Rev Macromol Chem C17:61
4. Lefebvre D, Jasse B, Monnerie L (1981) Polymer 22:1616
5. Fajolle R, Tassin JF, Sergot P, Pambrun C, Monnerie L (1983) Polymer 24:379
6. Lefebvre D, Jasse B, Monnerie L (1983) Polymer 24:1240
7. Tassin JF, Monnerie L (1988) Macromolecules 21:1846
8. Tassin JF, Monnerie L, Fetters LJ (1988) Macromolecules 21:2404
9. Doi M, Edwards SF (1978) J Chem Soc, Faraday Trans 2, 74:1789, 1802, 1818
10. Doi M, Edwards SF (1979) J Chem Soc, Faraday Trans 2, 75:32
11. Viovy JL, Monnerie L, Tassin JF (1983) J Polym Sci Polym Phys Ed 21:2427
12. Doi M (1981) J Polym Sci, Polym Lett Ed 19:265
13. Doi M (1983) J Polym Sci, Polym Phys Ed 21:667
14. Viovy JL (1986) J Polym Sci, Polym Phys Ed 24:1611
15. Tassin JF, Baschwitz A, Moise JY, Monnerie L (1990) Macromolecules 23:1879
16. Doi M, Pearson DS, Kornfield JA, Fuller GG (1989) Macromolecules 22:1488
17. Zhao Y, Jasse B, Monnerie L (1986) Makromol Chem, Macromol Symp 5:87
18. Zhao Y, Jasse B, Monnerie L (1990) Polymer 31:395
19. Berry GC, Fox TG (1968) Adv Polym Sci 5:261
20. Ferry J (1980) "Viscoelastic Properties of Polymers" 3rd Edition Wiley, New York
21. Oultache AK (1992) PhD Thesis, Université Pierre et Marie Curie, Paris
22. Wu S (1989) J Polym Sci B, Polym Phys Ed 27:723
23. Lauprêtre F, Bokobza L, Monnerie L (1993) Polymer 34:468
24. Wang CB, Cooper SL (1984) Adv Chem Ser 206:111
25. Bouton C, Arrondel V, Rey V, Sergot P, Manguin JL, Jasse B, Monnerie L (1989) Polymer 30:1414
26. Lefebvre D, Jasse B, Monnerie L (1984) Polymer 25:318
27. Wu S (1987) J Polym Sci, Polym Phys 25:2511
28. Lefebvre D, Jasse B, Monnerie L (1982) Polymer 23:706
29. Faivre JP, Jasse B, Monnerie L (1985) Polymer 26:879
30. Faivre JP, Xu Z, Halary JL, Jasse B, Monnerie L (1987) Polymer 28:1881
31. Zhao Y, Jasse B, Monnerie L (1989) Polymer 30:1643
32. Leonard C, Halary JL, Monnerie L (1985) Polymer 26:1507
33. Zhao Y, Jasse B, Monnerie L (1991) Polymer 32:209
34. Schmitt BJ, Kirste RG, Jelenic J (1980) Makromol Chem 181:1655



Authors' address:

Prof. L. Monnerie
ESPCI
10, rue Vauquelin
F-75005 Paris, France

**Progress in Colloid & Polymer Science** Progr Colloid Polym Sci 92:23—31 (1993)

# Lamellar morphologies in uniaxially-drawn banded spherulites of polyethylene

D. C. Bassett and A. M. Freedman

JJ Thomson Physical Laboratory, University of Reading, Reading, United Kingdom

*Abstract:* A morphological study, using permanganic etching and with lamellar resolution, of the uniaxial drawing at 80 °C of linear polyethylene containing banded spherulites has shown the response of the system to increasing draw ratios. Circular bands transform to increasingly eccentric ellipses to 30% extension but thereafter extend less than for affine change. Inhomogeneity in deformation develops with maximum response towards the ends of radii aligned along the draw direction and least for orthogonal radii. Dominant lamellae, which are initially S-profiled and non-planar flatten and their planes move towards the draw direction. Annealing experiments show that after drawing subsidiary lamellae melt and recrystallize at lower temperatures than dominant lamellae. This demonstrates that subsidiary lamellae are the more deformed and that there is a consequent spatial variation of mechanical properties typically of order 1 μm, the inter-dominant spacing.

*Key words:* Lamellae — morphology — spherulites — polyethylene — uniaxial drawing

## 1. Introduction

Structure-property relationships are of paramount interest to studies of polymeric materials and knowledge of them has been much advanced by the ability to study representative lamellar organization within a range of crystalline polymers [1]. This ability has progressively improved since the introduction of chlorosulphonation of polyethylene [2] and the advent of permanganic etching techniques applicable to various systems [3, 4]. In addition, we have demonstrated in earlier work how studies of representative lamellar organization can be supplemented by information on how molecular species are distributed systematically within the morphology [5, 6]. Such work has mostly been concerned with unoriented samples. There is much interest in seeing to what extent a similar approach may be applied to oriented sytems in which polymeric materials display some of their most interesting properties. In this paper we explore the application of permanganic etching to the development of orientation from isotropic polymer.

A priori, the prospects of revealing texture by etching are reduced in oriented, fibrous systems because the nature of lamellae and interlamellar regions is less differentiated whereas permanganic etchants penetrate and attack interlamellar material preferentially. Nevertheless, that there is texture showing shish-kebab morphologies was revealed in fibrous polyethylene long ago following treatment with fuming nitric acid [7]. Indeed, the fact that this reagent was too severe for unoriented polyethylene — which it rendered friable [8] — was a starting point for the introduction of permanganic etchants as a milder alternative. Permanganic etchants suitable for fibrous systems must, therefore, be made more fierce which, in general, means more aqueous. Examples have recently been published showing how fibres of different grades of polyethylene can be differentiated [9]. Other studies have, inter alia, revealed the presence of microscopic defective regions within commercial high modulus polyethylene fibres and details of how they can be bonded together to form macroscopic samples with good mechanical properties [10].

The work described in this paper is concerned with the formation of oriented materials by uniaxial tensile drawing. Its motivation was partly to explore how far existing etching techniques could continue to provide useful information as deformation proceeded and partly to explore within bulk polymer, with lamellar detail, phenomena of the kind first studied by Hay and Keller in thin films using optical and x-ray means [11]. Both objectives have been achieved. The conditions when deformation of banded spherulites departs from affine conditions have been established while systematic differences in the way different components of the morphology contribute to inhomogeneous deformation have been shown at both lamellar and spherulitic levels.

## 2. Materials and techniques

The linear polyethylene Sclair 2907 (du Pont of Canada) was used for all experiments. It was moulded into sheets, 1 mm thick by melting pellets between copper plates in a press, applying a sufficient load to squeeze molten polymer to the edges of the plates, then removing the load and quenching plates and polymer in cold water.

Tensile drawing was carried out with an Instron model 1112, mostly at 80 °C and a rate of extension of 0.05 cm $min^{-1}$ the slowest available speed. The dumbbell specimens used had a gauge length of 2 cm, a width of 4 mm and a thickness of 1 mm. These drawing conditions avoided the sharp neck given by fast strain rates but spread the neck region considerably, thereby facilitating examination of structure as a function of extension. The strain was measured directly on the sample from the distance apart of blank bars in a discontinuous thin carbon film previously evaporated on the surface of the sample through a flat copper gauze aligned along the gauge length.

Certain samples were subject to annealing at high pressure ($5.3_5$ kbar) as discussed later. Annealing prior to deformation was designed to increase lamellar thickness within the spherulites. Annealing after deformation was to emphasize phenomena of differential melting. These experiments were carried out in the high pressure apparatus described by Bassett & Turner, with samples immersed in silicone oil [12]. The chosen annealing temperature is approached slowly from below, then maintained within ± 0.05 °C for 0.2 h before cooling, at an initial rate of *c* 2K $min^{-1}$.

The principal means of examination has been transmission electron microscopy of carbon replicas of permanganically-etched surfaces. For this purpose samples were carefully cut in suitable directions into pieces approximately 2 mm square in cross-section then frozen on a specimen holder in a small pool of Tryco-m-bed (Aerosol Marketing and Chemical Co Ltd) using a hollow block containing dry ice. A rotary retracting microtome, (model 5030, Bright Instrument Co Ltd) with a glass knife, usually set at 15°, was then able to pare the specimen, in steps of 15 μm until all the sample was revealed through the Tryco-m-bed. At this stage the glass knife was replaced with a new one and further cuts of 5 μm taken from the surface of the specimen. The final exposed surface, which was cut parallel to a chosen direction, is the one used for etching, once the sample has been removed from the Tryco-m-bed, washed thoroughly first with distilled water, then methanol and dried.

For this work the permanganic etchant was prepared by dissolving crystals of potassium permanganate in a 2:1 mixture of concentrated sulphuric and dry orthophosphoric acids to a concentration of 0.07% w/v. The sulphuric acid was Analar grade (98%) while the dry orthophosphoric acid was prepared from as received stock boiled until a boiling point of 250 °C was attained. Etching was for about 1 h at room temperature after which the washing procedure described previously [3] was precisely followed.

Etched surfaces were usually replicated with cellulose acetate via a standard two-stage process. Exceptionally, when electron diffraction patterns were sought, direct replication with polyacrylic acid was attempted with the aim of detaching a portion of the polymer with the replica. This procedure was most successful when the specimen was immersed for several days in fuming nitric acid at room temperature, prior to permanganic etching. The result was to render the etched surface friable and so to leave portions of lamellae (mostly those flat on in the etched surface) attached to the replica. Such specimens allowed electron diffraction patterns to be recorded.

## 3. Results

### *3.1 Samples crystallized at atmospheric pressure*

The linear polyethylene of these experiments was chosen because it has a texture of banded spherulites whose bands serve as convenient markers for the changes occurring during deformation. If it is assumed that the bands in such spherulites are circular and sections of spheres — which is approximately but not strictly true — then in an affine deformation each sphere will transform into an ellipse with its major axis along the draw direction. For draw ratio $\lambda$, a sphere initially of radius $a$ will become an ellipsoid of revolution about the draw direction with major axis of length $\lambda a$ and equal minor axes of length $\lambda^{-1/2}\, a$.

In practice, deformation becomes inhomogeneous with increasing draw ratio as Figure 1 and similar data reveal. On the macroscopic level, an affine deformation implies firstly that, for a given band, the ratio of its major to minor axes varies as $\lambda^{3/2}$; the ratio of band periods, parallel and perpendicular to the draw direction will then also vary as $\lambda^{3/2}$. Secondly, for a diametral section (which may be selected in practice by choosing the largest number of visible bands) the minor axis and the band period along it will be a multiple $\lambda^{-1/2}$ of the corresponding dimensions in the undeformed spherulite. Figure 2 shows data plotted so as to test conformity with these predictions. In Fig. 2a the ratio of major to minor axes, each measured on the same band of a spherulite, is shown as a function of (draw ratio)$^{3/2}$. The data points fall increasingly away from the line of slope unity appropriate to affine deformation, as draw ratio increases beyond

Fig. 1. Bands revealed after drawing to ratios (a) 1 ie undrawn (b) 1.30 (c) 1.43 and (d) 2.20. In (b) to (d) the draw direction is parallel to the greatest extension of the bands. Note the apparent loss of detailed relief, especially in d as drawn morphologies are progressively less sensitive to the etchant. Bar equals 1 μm

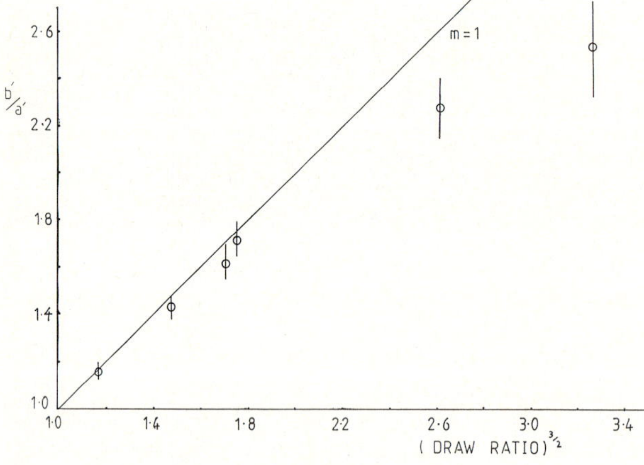


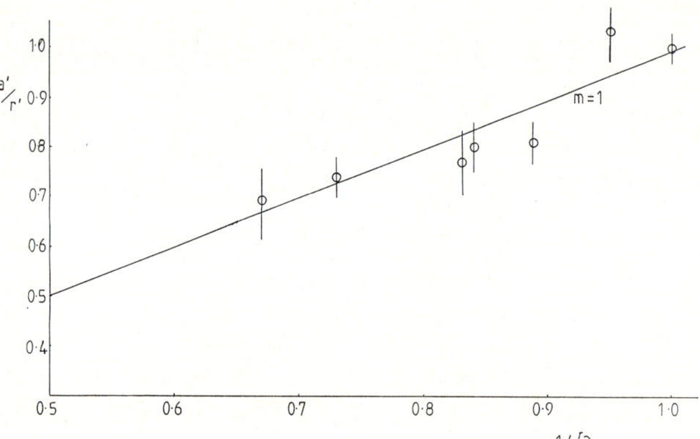


Fig. 2. Comparisons of band shapes with predictions for an affine transformation (full lines) 2(a): the ratio of major to minor axes for the same band as a function of $\lambda^{-1/2}$

about 1.3. On the other hand, in Fig. 2b the data for the ratio of the minor axis to the original ring diameter do fit well to the theoretical relationship with (draw ratio)$^{-1/2}$. The implication is that sperulites are contracting laterally as expected but failing to change shape parallel to the draw direction as an affine transformation would predict. The major axis is less extended than anticipated so that, as there is no indication of strain being accommodated interspherulitically, it must follow that, to conserve volume, deformed spherulites have adopted a shape tending towards the cylindrical and increasingly differing from the uniaxial ellipsoid appropriate to affine deformation as may be seen in Fig. 1. The reasons for this must concern the details of microstructure. Indeed, it is evident that even when gross deformation appears to be affine the spherulitic radii after drawing no longer diverge from the centre of the ellipse but from two points to either side (Fig. 3).

Fig. 3. In a spherulite drawn 1.3 times, the microstructure no longer appears to radiate from a common centre. Bar equals 1 μm

Fig. 4. A comparison of lamellae in bands for radii (a) parallel and (b) normal to the draw direction, in the same deformed spherulite. Bar equals 1 μm

Increasingly at higher draw ratios, the bands reveal inhomogeneous deformation. There is also a slower rate of etching so that the more drawn regions can be seen to stand comparatively high when surface relief of etched samples is inspected by Nomarski differential interference contrast optics. Differential relief between lamellae and their surroundings is also reduced (Figs. 1d, 8) presumably because of decreased penetration of the etchant into interlamellar regions.

Inhomogeneity becomes evident particularly in two ways. One is that, along the major axis, tips of spherulites and their bands become sharp and not rounded. This is the case in Fig. 4a. For comparison, Fig. 4b shows the region of the minor axes of bands in the same deformed spherulite, i.e. where bands lie parallel to the draw direction. Notice in both locations that lamellae have changed profile and become flat or flatter. The second aspect of inhomogeneity is perturbation of banding with ripples developing on what, for affine deformation, would be elliptical contours. This can eventually lead to the phenomenon shown in Fig. 5 where a sector close to, but not necessarily along, the draw direction is differentiated from the remainder of the spherulite and appears to take on a separate identity. Such features are particularly prevalent for deformation at room temperature and may well be analogous to the caps of comparatively undistorted banded spherulite within a highly deformed matrix reported by Hay & Keller [11].

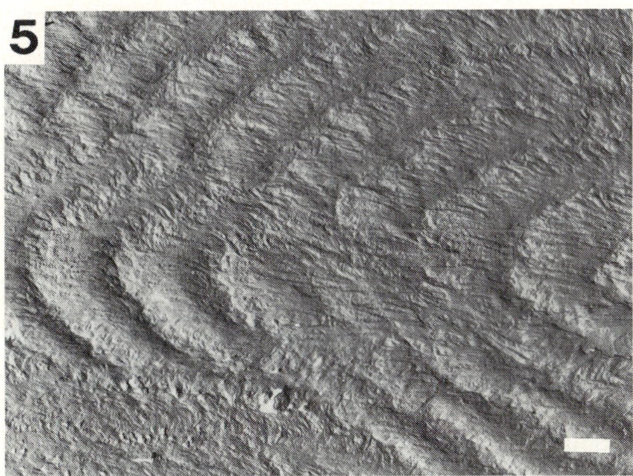


Fig. 5. In a sample drawn at room temperature a segment of the spherulite is distinguished by being less deformed than its surroundings. Bar equals 1 μm

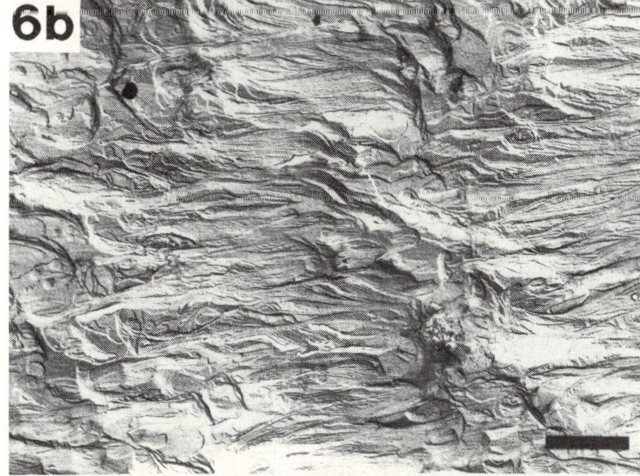


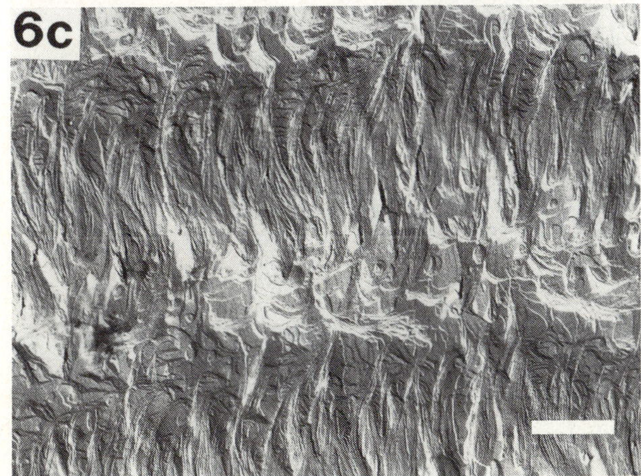


Fig. 6. A comparison of lamellae in a sample annealed at high pressure prior to drawing at 80 °C (a) undrawn bands (b) normal and (c) parallel to the draw direction in the same spherulite. Bar equals 1 μm

Fig. 7. A view down the draw direction of a pre-annealed spherulite within an extended neck region after drawing at 80 °C shows lamellae which are predominantly planar with normals lying in a cone around the draw direction. Bar equals 1 μm

Fig. 8. The structure within highly-drawn spherulites is less readily etched but does show lamellae lying along the draw direction while still retaining traces of the original banding. Bar equals 1 μm

The inhomogeneity of deformation is also demonstrated by electron diffraction. For example, a banded spherulite with rings intact after being extended 2.2 times showed single crystal patterns in the waist showing that the *b* axis remains transverse to the draw direction. In the tip regions, however, *c* axis alignment is indicated by fibre patterns with hkO reflections equatorial to the draw direction.

### 3.2 Samples preannealed at high pressure

Annealing banded spherulites of linear polyethylene at high pressure increases the thickness of lamellae at constant orientation i.e. with preservation of the band period [1, 16]. By this means annealing at $5.3_5$ kbar and 230 °C has promoted better resolution of detail in deformed and etched samples.

Figure 6a is of an undeformed spherulite prepared by pressure annealing as described above. In a slightly deformed spherulite, bands normal to the draw direction (Fig. 6b) show that lamellae, while preserving some hierarchical substructures, have become predominantly planar and approached parallel orientations with the tensile axis: the curved lamellae present in Fig. 6a are now largely absent. By contrast, in bands parallel to the draw direction in the same spherulite (Fig. 6c) lamellae are clearly curved with edge-on profiles being nonlinear and C or S-shaped. Certain dominant lamellae can be seen, especially when more or less flat on, to have become disrupted by shear approximately in {020} planes, leaving traces parallel to the draw direction. A view of such a sample looking down the draw direction, obtained by cutting the sample open across the neck, is shown in Fig. 7. It, too, shows how lamellar surfaces have begun to move towards parallelism with the draw direction. At higher levels of deformation the texture revealed by this etchant (Fig. 8) is substantially one of parallel sheets but still shows traces of the initial banding.

### 3.3 Samples annealed after drawing

The changes caused by deformation will reduce the melting point of lamellae, the more so the greater the associated increase of Gibbs free energy. Inhomogeneous deformation will, therefore, be reflected in a spatial variation of melting point. It is possible to reveal this by suitable choice of annealing conditions. Figure 9 illustrates this effect for an-

Fig. 9. Differing melting points revealed by isothermal annealing at high pressure following deformation at 80 °C. (a) and (b) at 230 °C (c) at 233 °C. The undrawn spherulite in (a) shows no sign of melting and crystallization whereas that in (b), from the neck region of the same sample, shows traces of recrystallized lamellae, normal to the draw direction, between the bands of surviving, flat-on dominant lamellae. In (c) more of the sample has melted and recrystallized together with bands of surviving dominant lamellae. Bar equals 1 μm

nealing at $5.3_5$ kbar and two temperatures. The spherulites in Figs. 9a and 9b are part of the same specimen annealed at 230 °C. That in Fig. 9a was undeformed and shows no differential effects of annealing. That in Fig. 9b is further into the neck region and shows evidence of having melted and recrystallized selectively in the region of their major axes. (Elsewhere it is seen that the original lamellae have survived but in a deformed state, revealing traces, when edge-on which would have been radial in the underformed spherulite.) Along major axes, however, the original lamellae have all but disappeared to be replaced by others with normals more or less parallel to the draw direction. In adjacent areas there are traces of differential melting with original dominants having survived but subsidiaries transformed (e.g. in the upper two bands at the top right of Fig. 9b).

At the higher annealing temperature of 233 °C much more of the sample has melted and recrystallized but original lamellae remain clearly evident in the bands which stand high in the etched surface. Elsewhere there has been extensive recrystallization. These recrystallized lamellae indicate by the direction of their normals the orientation of the nuclei on which they have recrystallized. This is principally along the draw direction but note the alternate ridging between the bands at the lower left which implies systematic and repetitive deviations away from the draw direction, whose origin remains to be clarified but, at least superficially, is similar to the banded textures linked to relaxation in liquid crystalline polymers.

## 4. Discussion

The development of orientation in semi-crystalline polymers has received a great deal of attention

[13, 14]. It is a complex subject, not least because of the textural complications present in the starting materials. Among studies on polyethylene which have attempted to progress by simplifying the initial morphologies are a long series by Geil et al. [15] and work of Allan & Bevis [16] both examining the response of individual solution-grown lamellae, stretched on a substrate; papers by Hay & Keller [11] on deforming banded spherulites and work by Attenburrow & Bassett, in which thick, pressure-crystallized lamellae were used both to monitor response of individual lamellae within the polymer during deformation [17] and to demonstrate that uniaxially drawn material contained one population of lamellae which had survived the draw ableit much deformed and a second of transformed lamellae [18]. With the advent of permanganic etching, these last cited studies have been taken considerably further by Hodge & Bassett (unpublished but see chapter 9 in reference [19]). The work in this paper has applied permanganic etching to uniaxially-drawn banded spherulites.

It shares with the work of Hay & Keller the advantage of having the bands of the spherulites as markers to identify stages of deformation but has two principal points of difference with that pioneering study. One is simply that of greatly improved resolution in the use of the electron microscope. The second is more profound: we are examining representative planes within the interior of thick deformed specimens. In consequence not only may spherulitic radii be substantially inclined to the plane of the paper but also, more importantly, the constraints on spherulites as they deform are quite different from those for thin-film specimens. In a thin film specimen, grown as such, all spherulitic radii lie in or close to its plane. Moreover, inter-spherulitic boundaries, being the last places to crystallize, may be thinner than the rest of the film and so be prone to yielding. It is noteworthy that such yielding, while commonly reported by Hay & Keller, has not occurred in our experiments.

In Fig. 1 and subsequent figures, one sees the residual plastic strain after deformation. These and other illustrations have been chosen because they show large numbers of bands so will be of approximately diametral sections of spherulites. According to the data of Fig. 2a, macroscopic (band) dimensions are consistent with an affine transformation to *c* 30% strain. Microscopic detail (Fig. 3), however, is already inconsistent with such a scheme. Inhomogeneity becomes evident in the differences between textures along major and minor axes of deformed spherulites. It is along the major axis that deformation is greater (Fig. 4a) with the development of a sharp tip and an associated fibre electron diffraction pattern. At the same time, Fig. 4b (and associated electron diffraction patterns) shows that deformation in the waist (minor axis) region can leave lamellae comparatively unaltered.

The presence of caps of little-deformed spherulitic sectors within a deformed matrix was reported by Hay & Keller [11] and may well be related to the phenomenon shown in Fig. 5 for a sample drawn at room temperature. The origin of this feature is very likely related to yielding along planes of maximum shear stress. In so far as a distinct sector is separated, the implication is that the yield criterion favours yielding along the boundaries of this sector as opposed to alternative processes. It is entirely reasonable that such a differentiation should be more prominent at room temperature than at 80 °C, as our results suggest, because of the softening of crystal moduli rising temperature will bring.

The effect of deformation on lamellae within the spherulite is most clearly evident in samples which have been pre-annealed at high pressure. Although high pressure is most often associated with transition to the disordered high-pressure phase and the formation of thick, anabaric lamellae [19], it must be emphasized that the conditions used here lie below the conditions previously monitored, in the same apparatus, for the first order transition to occur [20]. All morphological evidence is also consistent with the polymer remaining orthorhombic at all times. Annealing under these conditions is known to produce lamellar thicknesses of *c* 50 nm while preserving the original band period [21, 22, 23].

As shown in Fig. 6, the greater deformation of the most-highly-drawn regions (i.e. near the tips of the major axes of deformed bands) has the effect of making lamellae in them more nearly planar whereas around the waist, lamellae have curved traces in planes normal to the draw direction. Elucidation of the precise crystallography of processes affecting lamellae requires more detailed electron diffraction evidence (as in refs. [15, 16]) in relation to the lamellae resolved than has so far proved possible. The results are, nevertheless, in aggreement with the earlier studies in this laboratory of Attenburrow & Bassett and Hodge & Bassett cited above, which have each shown that, with increasing draw ratio, lamellar planes and molecular *c* axes both move towards the draw direction but from opposite sides. The morphology of Fig. 7 is consistent with this finding in that it shows lamellar planes lying in a cone around the draw direction. Highly drawn regions (Fig. 8) have a texture of lamellae aligned along the draw direction but traces of the original spherulitic organization are visible as coarse relief.

The extent of deformation, as measured by the reduction in melting point is brought out in Fig. 9. The greater deformation along the major axis of a deformed band is revealed by the recrystallization produced in these regions (Fig. 9b) whereas no such effect is apparent along the minor axis. A higher annealing temperature (Fig. 9c) produces a more extensive recrystallization. In interpreting such morphologies one must bear in mind that the location of a band in a spherulite is a function of the viewing direction. The average orientation of lamellae spirals around the radial *b* direction in undeformed spherulites (although individual lamellae change orientation discontinuously around giant screw dislocations). Thus, in Fig. 9b one can see, at the upper right especially, lamellae edge-on whose traces are radial in the drawn spherulite; these have survived the draw, albeit in a deformed state. In between close scrutiny shows reformed lamellae normal to the draw direction.

One concludes, therefore, that the surviving lamellae are the dominants while those melting and recrystallizing are the subsidiary lamellae lying between them. Mechanical properties are thus varying on the scale of the interdominant separation. This finding reproduces in an equivalent context, the conclusion drawn previously in our studies of quantitative permanganic etching [5].

## Conclusion

A study, with lamellar resolution, of the morphology of banded spherulites of polyethylene, uniaxially drawn at 80°C has led to the following conclusions.

1) Circular bands transform to increasingly eccentric ellipses to 30% extension but thereafter extend less than for an affine change and show inhomogeneous response. Even at 30% extension, the internal microstructure shows the deformation not to be strictly affine.
2) Greatest deformation occurs towards the outside of the spherulite along radii parallel to the draw direction; orthogonal radii show least change at low draw ratios. There is no evidence of interspherulitic yielding and the presence of comparatively undeformed segments or caps within a more-highly-drawn matrix is much less than for drawing at room temperature.
3) In responding to increasing draw ratio dominant lamellae, which initially are S-profiled and nonplanar, flatten and their planes move towards the draw direction.
4) Subsidiary lamellae suffer greater change on drawing than do dominants and hence show a more reduced melting point. The differential response to isothermal annealing reveals a spatial variation of mechanical properties of the inter-dominant separation i.e. of around 1 μm.

*Acknowledgement*

AMF is indebted to the SERC for a postgraduate studentship.

## References

1. Bassett DC (1988) In: Bassett DC (ed) Developments in Crystalline Polymer-2, Elsevier Applied Science, London, pp 67 111
2. Kanig G (1973) Kolloid Z 251:782
3. Olley RH, Hodge AM, Bassett DC (1979) J Polym Sci, Polym Phys Ed 17:627—643
4. Olley RH, Bassett DC (1982) Polymer 23:1707
5. Freedman AM, Bassett DC, Vaughan AS, Olley RH (1986) Polymer 27:1163
6. Freedman AM, Bassett DC, Olley RH (1988) J Macromol Sci-Phys B27:319—335
7. Ingram P, Peterlin A (1964) J Polym Sci B2:739
8. Keller A, Sawada S (1964) Makromol. Chem 74:190
9. Tsui S-W, Duckett RA, Ward IM, Bassett DC, Olley RH, Vaughan AS (1992) Polymer 33:4527—4532
10. Olley RH, Bassett DC, Hine PJ, Ward IM (1993) J Mat Sci 28:1107
11. Hay IL, Keller A (1965) Kolloid Z 204:43
12. Bassett DC, Turner B (1973) Phil Mag 29:285
13. Peterlin A (1971) J Mat Sci 6:490
14. Stein RS (1969) Polymer Eng Sci 9:320
15. Kiho H, Peterlin A, Geil PH (1965) J Polym Sci B3:263
16. Allan P, Bevis M (1974) Proc Roy Soc A341:75
17. Attenburrow GE, Bassett DC (1979) Polymer 20:1313
18. Attenburrow GE, Bassett DC (1977) J Mat Sci 12:192
19. Bassett DC (1981) Principles of Polymer Morphology, Cambridge University Press, Cambridge
20. Bassett DC, Turner B (1973) Phil Mag 29:925
21. Rees DV, Bassett DC (1968) Nature 219:368—370
22. Bedborough DS (1974) PhD Thesis, University of Reading
23. Bassett DC, Khalifa BA, Olley RH (1976) Polymer 17:284—290

Authors' address:

Prof. D. C. Bassett
JJ Thomson Physical Laboratory
University of Reading
PO Box 220
Whiteknights Reading RG6 2AF, United Kingdom

# Formation of highly oriented films by epitaxial crystallization on polymeric substrates

J. C. Wittmann*), B. Lotz*), and P. Smith**)

*) Institut Charles Sadron, Strasbourg
**) Materials Department, University of California of Santa Barbara, Santa Barbara, California, USA

*Abstract:* Polymeric substrates with widely varying degrees of orientation and perfection have long been used to align liquid-crystals or to epitaxially orient crystalline compounds including organics or other polymers. In the latter case and as with low molecular weight substrates, epitaxy results in the formation of highly structured polymer films made of lamellae standing edge-on. However, it is only with single crystal-like polymeric substrates that the underlying crystallographic relationships can be unequivocally determined and/or the relevant physical film properties really optimized. The technique of friction transfer of PTFE developed recently leads to large organized surfaces and provides a powerful and versatile method to orient a wide range of organic materials.

*Key words:* Epitaxy — orientation — polymeric substrates — PTFE

## Introduction

Epitaxial growth of crystalline polymers results in the formation of highly structured layers made of lamellae standing edge-on, i.e., with the molecular axes oriented in the layer plane [1, 2]. In the most favorable cases (mostly encountered with crystalline substrates of low surface symmetry) a unique, single crystal-like orientation of the polymer may be achieved [3], whereas polymer samples oriented by more classical means usually have a fiber-like orientation. The epitaxial growth technique is well-suited for the preparation of samples adequate for structural and morphological investigations and offers also a means of optimizing the physicochemical or physical properties of the deposited layers in view of specific applications. Interactive substrates used so far for polymer epitaxy belong mainly to three different categories: i) alkali halides and other inorganic substrates such as mica or quartz [1, 2], ii) organic compounds, e.g., condensed aromatic hydrocarbons, linear polyphenyls or aromatic acids and salts [3], and iii) polymer films or layers oriented by various techniques including unidirectional rubbing [4, 5], uniaxial drawing [6, 7], gel spinning [8] or friction transfer [9].

The present contribution reviews the current status of epitaxy or alignment using polymeric substrates and describes very briefly the formation, properties and uses of friction deposited polytetrafluoroethylene (PTFE) layers. As will become apparent, epitaxial growth on polymeric substrates may offer interesting alternatives to more conventional orientation methods such as the Langmuir-Blodgett technique.

Polymeric substrates have long been known to be able to nucleate and orient the growth of a wide range of materials including crystalline polymers, liquid crystalline compounds or even metals. Actually, substrates ranging from bulk crystallized samples, uniaxially drawn films or fibers to quasi monocrystalline layers, i.e. with widely varying degrees of orientation and perfection, have been used in numerous studies on polymer-polymer crystallographic interactions. With regards to the more specific case of liquid-crystalline compounds to be discussed first, very thin polymer films rubbed or buffed in a single direction have been privileged, essentially for practical reasons.

## I. Rubbed polymer layers

Homogeneous alignment of nematic or smectic liquid-crystal materials or of their mixtures on thin, rubbed polymer layers is a well established method. It has recently gained renewed interest with the discovery of surface stabilized ferroelectric L. C. displays (SSFLC) [10]. Usually, a thin polymer coating is deposited from solution on a glass support and, after complete drying, gently rubbed in a single direction with a suitable tissue. This process is thought to induce the formation of grooves and to orient the polymer chain at or near the surface in the rubbing direction. It has been claimed that a crystalline polymer is required for optimal alignment [11, 12]. The detailed alignment mechanism is not yet fully understood. As has been suggested by Geary et al. [11] this mechanism may be analogous to the epitaxial growth of conventional crystalline solids. This conclusion is supported by the work of Myrvold [12, 13], who examined over 100 different polymer layers and pointed out that even the crystal structure of the layer is of importance, since monoclinic or triclinic structures appear to be more efficient than structures of higher symmetry for successful alignment in SSFLC cells.

Recently, the use of rubbed polymer layers has been extended to the oriented growth of polydiacetylene thin films in order to produce large samples and try to maximize their non-linear optical properties [14—16]. Such films are readily prepared by epitaxial polymerization, i.e., solid-state photopolymerization of the diacetylene monomers which have been vapor deposited under vacuum and pre-oriented on the rubbed layers (polyesters or polydiacetylene). Birefringence measurements indicate that fairly high degrees of orientation are achieved.

In both fields, lack of knowledge on the exact orientation mechanism limits further developments and more structural studies are needed to fully elucidate the orientation mechanisms and interactions at play.

## II. Mechanically drawn polymer films

Following the early discovery of Willems in 1963 [17] on the oriented growth of polyethylene and paraffins on drawn polyoxymethylene films, numerous investigations on epitaxial interactions between two different polymer species have since been undertaken. These investigations were mostly aimed at understanding the properties of fiber reinforced materials or elucidating the crystallographic interactions which lead to enhanced nucleation or improved mechanical properties in polymer-polymer blends. Perfect macroscopic single crystals being unaivalable, uniaxially melt-drawn polymeric substrates were extensively used instead. Petermann and Gohil [6] have devised an ingenious melt-draw technique leading to oriented polymer films with thicknesses in the 50 nm range. The structure of these films is made of a central fibrous core overlaid with a dense array of parallel, chain-folded lamellae standing edge-on. Very useful for the formation of thin polymeric substrates of large size, the method has also been successfully applied to the orientation of polymer blends or the formation of polymer multilayers or laminates [18, 19].

The attainable level of orientation of the deposited species depends drastically on the substrate structure and perfection. The major advantages of drawn polymer films are ease of preparation and practically unlimited size. However, they possess per essence a fiber structure characterized by a unique chain axis orientation but with various crystallographic planes exposed. As a consequence, the exact epitaxial relationship between the polymer species can hardly be determined.

A screening of a variety of uniaxially drawn crystalline polymer substrates including polyolefins, polyamides and polytetrafluoroethylene (PTFE) was performed by Takahashi and coworkers [20, 21] and Willems [22] in the 1970s. Their findings are summarized in Table 1 (taken from ref. [3]). Note that some negative results, especially those concerning isotactic polypropylene (iPP) are contradicted by more recent findings. This is probably due to ill-adapted growth procedures or poorly oriented substrate films.

More recently, Petermann and coworkers have concentrated on bilayers, multilayers or blends of highly oriented iPP, a substrate with a $3_1$ helical chain conformation, with a second polyolefine-type component including polyethylene (and paraffins) [7, 23], polybutene-1 [24], polyoctenamer [25] or trans 1,4-polybutadiene [26]. Thanks to the preparation and observation methods used by these authors, more detailed information could be gained on the polymer/substrate reciprocal orientations. With the exception of polybutene-1 for which no epitaxy was observed, the other polymers with a zigzag chain conformation orient epitaxially on iPP with their chain axes inclined at ±50° to the iPP draw or chain direction. This gives rise to typical "cross-hatched" lamellar morphologies. The latter are thought to be responsible for the synergetic effects in mechanical properties observed with iPP/PE multilayers or blends [18].

Table 1. Polymer/polymer epitaxies determined by Takahashi et al.

| Substrate | POM (1) | PA (1) | PE (1) | i-PP (1) | PTFE (2) |
|---|---|---|---|---|---|
| Deposit | | | | | |
| PEA | n.d. | − | + | n.d. | n.d. |
| PCL | + | + | + | − | + |
| PA | n.d. | | + | n.d. | + |
| PE | + | − (1)<br>+ (3) | | − | + |
| i-PP | n.d. | + | n.d. | | − |

1) T. Takahashi et al. J. Polym. Sci., Polym. Lett. Ed. 8, 651 (1970).
2) T. Takahashi et al. J. Macromol. Sci.-Phys. B12 (3), 303 (1976).
3) J. Willems Kolloid-Z. u. Z. Polym. 251, 496 (1973).

+ : epitaxy  − : absence of epitaxy
n.d. : not determined

Finally, according to the experimental evidences presented, it appears that the epitaxial relationships between the substrate of helical iPP and the various zigzag chains are based on a common molecular mechanism, a point of view which is somewhat challenged by Petermann and coworkers [23]. As will be described next, the details of the underlying epitaxial growth mechanism could only be worked out by resorting to single crystal-like polymeric substrates [27].

## III. "Monocrystalline" polymeric substrates

Only this third category of polymeric substrates can yield deposited layers with the high degree of perfection and orientation needed either to determine unequivocally the underlying epitaxial relationships or, more importantly to optimize some relevant physical property. In this connection, use has been made of i) epitaxially oriented layers of polymers on low molecular weight organic substrates, ii) high modulus fibers or films of polyethylene prepared by the gel spinning method and iii) friction transferred PTFE layers.

### *III.1) Epitaxially grown polymer layers*

The clue for the orientation mechanism of linear zigzag chains (PE or polyamides) on iPP was obtained by carefully investigating the two following well defined systems in which iPP is alternatively the substrate or the deposit [27, 3]:

- iPP quadrites grown from solution and decorated with condensed PE vapors,
- iPP lamellar crystals grown and annealed on a thin polyamide 6 single crystal-like layer itself produced beforehand by epitaxial growth on benzoic acid.

The detailed epitaxial relationship deduced from the composite diffraction patterns of these bilayers led us to propose the structural scheme shown in Fig. 1. In brief, this scheme is based on the existence in the iPP contact plane of characteristic rows of methyl (Me) groups and the parallel alignment at the polymer/polymer interface of these rows and the linear zigzag chains. Thus, to a unique orientation of the latter correspond two orientations (at ±50°) of the iPP helices (and therefore also of iPP lamellae) according to their handedness in the contact plane. Last but not least, this scheme implies also a close match between the Me interrow distance and the PE or PA6 interchain distance in the contact plane [27].

Extension of this structural scheme to other interacting polymer systems involving namely syndiotactic polypropylene is currently under investigation [28, 29].

### *III.2) Ultra-drawn polyethylene film*

A second alternative for large and perfectly oriented substrates is offered by ultra-high molecular weight PE thin films drawn from the gel state to draw ratios $\lambda > 100$. A double orientation is obtained with the PE chain axis perfectly oriented in the drawing direction and the unit-cell a and b axes perpendicular and parallel to the film plane, respectively. In contradistinction to melt-drawn films, such ultra-drawn PE films thus expose a well defined (100) crystallographic contact plane [30].

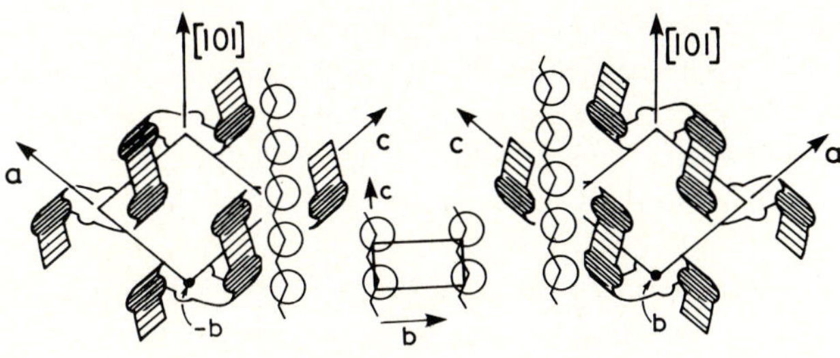


Fig. 1. Schematic representation of the relative orientations of PE chains and right- or left-handed iPP helices in the (100) and (010) contact planes, respectively. The rows of Me groups of iPP which are represented by ellipses, as well as the PE chains are vertical, i.e., at an angle of 50° to the two different iPP chain axes

An attempt was made to prepare thin, oriented films of the polar β phase of poly(vinylidene fluoride) ($PVF_2$) by vapor deposition on PE substrates. Indeed, it has been claimed that $PVF_2$ films in the β form could be prepared directly by vapor deposition on silicon wafers held at fairly low temperature [31]. Furthermore, the β form bears a close structural similarity with PE (all trans). However, our attempts to induce the oriented growth of this polar phase failed, since the predominant phase was the apolar α [32].

Using the same substrate, Kawaguchi and coworkers [33, 34] have epitaxially oriented both long- and short-chain compounds. The dominant epitaxial orientation of iPP on ultra-drawn PE is very similar to that described above for a polyamide substrate. Two additional orientations in minor proportions were also observed [33]. Series of short chain compounds comprising n-paraffins, n-alcohols and n-carboxylic acids were also successfully oriented on PE substrates with well characterized contact planes and a (short) chain to (long) chain alignment was revealed [34].

Ultra-drawn PE films offer an interesting route for the preparation of thin, highly oriented layers of various materials. Unfortunately, their use is severely restricted by the rather low melting temperature of this polyolefin.

*III.3) Friction transferred poly(tetrafluoroethylene) layers*

In a series of experiments performed some years ago, Tabor and collaborators [35, 36] have thoroughly studied the frictional behavior of PTFE sliding over a clean smooth surface of metal or glass. They showed that in the low friction régime (proper to PTFE and PE), i.e., at low sliding speeds and/or moderately high temperatures, a thin (in the 10 nm range), adherent PTFE layer is continuously transferred on the counter-surface. According to the authors, this process has its origin in the rather smooth molecular profile of the PTFE chains [36].

As demonstrated more recently by two of us [9], such friction-deposited layers may act as very efficient substrates for the oriented growth of a wide variety of materials including liquid crystals [37], organic compounds and, of course, liquid-crystalline or crystalline polymers. In accordance with the previous observations [36] electron microscopy, electron diffraction [9] as well as atomic force microscopy investigations [38] clearly show that the PTFE layers do not have the lamellar overgrowth described previously for the drawn polyolefin films. However, they are not molecularly smooth since many steps of various heights (1—30 nm) extend along the sliding path, i.e., parallel to the PTFE chain axis. The structure of these layers (Figs. 2a, b) can nevertheless be termed as "quasi monocrystalline" and the high degree of orientation and crystalline perfection reached compares favorably with that of extended chain PTFE wiskers formed directly by synthesis [39] or of gel processed PE films [30].

Use of friction deposited layers offers several important advantages:

- ease of preparation over fairly large surfaces (several $cm^2$) on various types of supports including as already mentioned, metals and glass but also Si wafers, ITO electrodes, etc.
- total transparency, a property which is of great interest for optical applications.
- high chemical as well as thermal stability (up to 300°C), thus offering the possibility to use corrosive solvents such as sulfuric acid or to submit the deposit to appropriate annealing treatments.

However, these layers suffer from a relative fragility and from a lack of wettability well known for fluorinated surfaces and which in some cases (deposition from solution), calls for special preparation techniques.

Many more materials than those listed in Table 1 of reference [9] have been successfully oriented on PTFE layers, but a detailed description of these results is out of the scope of the present paper. By way of illustration, emphasis is currently put on a series of investigations on deposition and orientation processes involving epitaxial polymerization, i.e., vaporization under vacuum of the monomer or its precursor, condensation on the PTFE layers and either simultaneous polymerization during condensation or subsequent polymerization in the solid state after condensation.

The system PTFE/poly(p-xylylene) (PPX) belongs to the first category since polymerization and orientation of the PPX crystalline α form are performed simultaneously by using the well known Gorham's method [40]. The monomer formed under vacuum by vaporization and pyrolisis at 600 °C of the chemically more stable dimer, is indeed directly condensed-polymerized on the PTFE layer held at room temperature. The as-deposited PPX film in the α form can subsequently be transformed into the more crystalline and in turn highly oriented PPX β form by short-time annealing at 300 °C. As demonstrated by Figs. 2c, d poly(p-xylylene) films elaborated by this method are uniform even after annealing and highly oriented with the PPX chain axis running parallel to the PTFE one.

Oriented polydiacetylene films on the contrary are best prepared by a two-step method already mentioned for rubbed polymer layers: oriented growth under vacuum of the monomer layer on the exposed PTFE surface, followed by photopolymerization in the oriented solid state. Once more, the composite electron diffraction patterns give

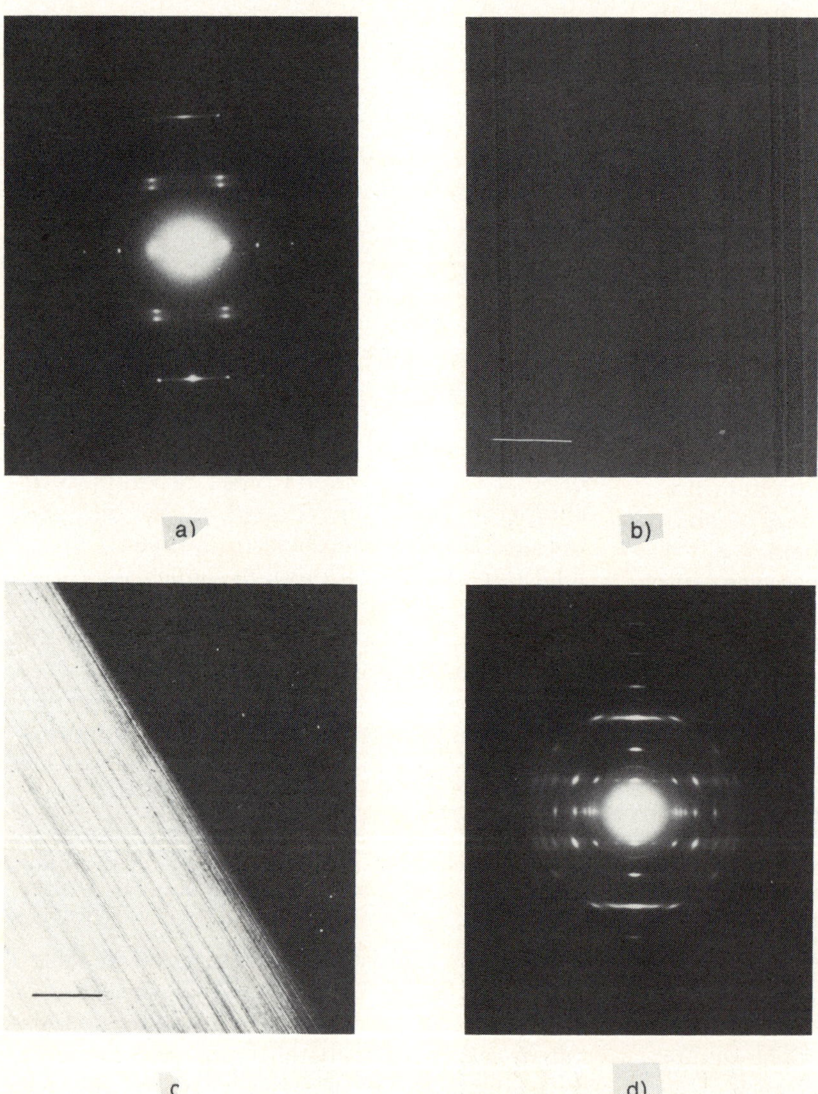


Fig. 2. a) Electron diffraction pattern of a friction transferred PTFE layer; chain axis vertical. b) Electron micrograph of a gold decorated PTFE layer; friction direction and PTFE chain direction vertical. Gold wets only the uncovered glass surface; scale bar: 0.15 μm. c) Thin poly(p-xylylene) film vapor deposited and polymerized on a glass slide partially covered with an oriented PTFE layer (right-hand-side). Note the high birefringence due to the orientation of the PPX film induced by PTFE. Optical micrograph with crossed polarizers; scale bar: 100 μm. d) Electron diffraction pattern of PPX oriented PTFE; PPX and PTFE chain axes vertical

evidence of a high degree of order in these "epitaxially polymerized" polydiacetylene thin films which are again characterized by a parallelism of the deposit/substrate chains and probably also by a lattice matching of the two crystalline species at the interface.

The fact that the PTFE chains run parallel to the friction direction and therefore also parallel to the numerous steps covering the contact surface complicates the analysis of the orientation mechanism(s) of PTFE friction deposited layers. Indeed, since all except one (iPP [41]) of the polymers studied so far have their chain axis oriented parallel to the PTFE chains, distinction between preferential adsorption, orientation and nucleation along the PTFE steps and/or lattice matching in the same or in a lateral direction is not easy. Evidence has been gained both for a very specific mechanism based on lattice matching and for a less specific one based on preferential nucleation along steps which has sometimes also been invoked for L. C. alignment on rubbed polymer layers. In search of an answer, experiments with PTFE layers covered with carbon coatings [42] are in progress. Whatever the answer, the method described is very simple and versatile and offers great potential in the many fields where ordering of materials used is mandatory.

## Conclusion

Epitaxy has often been claimed as being inappropriate to produce large and well organized films of organic materials, including polymers. The reasons recurrently put forward are mainly a lack of generality of the phenomenon (or else its high specificity) and the unavailability of large, efficient and versatile substrates able to induce the unidirectional growth of the deposit. As shown in the present paper, the situation is rapidly changing with the emergence of uniaxially-drawn thin films or quasi-monocrystalline polymeric substrates such as ultra-drawn polyethylene or friction transferred PTFE layers. Such substrates offer significant advantages over inorganic crystals which, due to their high symmetry, very often lead to multiple orientations or organic substrates which, although very efficient, are seldom available as macroscopic single crystals. The thin macroscopic and highly oriented polymeric substrates offer powerful new routes for tailored growth of crystalline polymers and of a wide range of short chain compounds or low molecular weight organic materials.

*Acknowelegment*

The authors wish to thank S. Meyer, S. Graff, F. Motamedi, and K. J. Ihn for their assistance.

## References

1. Mauritz KA, Baer E, Hopfinger AJ (1978) J Polymer Sci Macromol Rev 13:1
2. Swei GS, Lando JB, Rickert SE, Mauritz KA (1986) Encyclopedia of Polymer Science and Engineering vol 6:209
3. Wittmann JC, Lotz B (1990) Prog Polym Sci 15:909
4. Cognard J (1982) Mol Cryst Liq Cryst Suppl 1:1
5. Patel JS, Leslie TM, Goodby JW (1984) Ferroelectrics 59:137
6. Petermann J, Gohil RM (1979) J Mater Sci 14:2260
7. Broza G, Rieck U, Kawaguchi A, Petermann J (1985) J Polym Sci Polym Phys Ed 23:2623—2627
8. Smith P, Lemstra PJ (1980) J Mater Sci 15:505
9. Wittmann JC, Smith P (1991) Nature 352:414
10. Clark NA, Lagerwall TS (1980) Appl Phys Lett 36:899
11. Geary JM, Goodby JW, Kmetz AR, Patel JS (1987) J Appl Phys 62:4100
12. Myrvold BO (1988) Liquid Cryst 3:1255
13. Myrvold BO (1991) Mol Cryst Liq Cryst 202:123
14. Kanetake T, Ishikawa K, Koda T (1987) Appl Phys Lett 51:1957
15. Patel JS, Lee Sin-Doo, Baker GL, Shelburne JA, III (1990) Appl Phys Lett 56:131
16. Kanetake T, Tomioka Y, Imazeki S, Taniguci Y (1992) J Appl Phys 72:938
17. Willems J (1963) Naturwissenschaften 50:92
18. Gross B, Petermann J (1984) J Mater Sci 19:105
19. Petermann J, Broza G, Rieck U, Kawaguchi A (1987) J Mater Sci 22:1477
20. Takahashi T, Inamura M, Tsujimoto I (1970) J Polym Sci Polym Lett Ed 8:651
21. Takahashi T, Teraoka F, Tsujimoto I (1976) J Macromol Sci Phys B12:303
22. Willems J (1973) Kolloid Z Z Polym 251:496
23. Petermann J, Xu Y (1991) J Mater Sci 26:1211
24. Jaballah A, Rieck U, Petermann J (1990) J Mater Sci 15:3105
25. Xu Y, Asano T, Petermann J (1990) J Mater Sci 25:983
26. Petermann J, Xu Y, Loos J (1992) Makromol Chem 193:611
27. Lotz B, Wittmann JC (1986) J Polym Sci, Polym Phys Ed 24:1559
28. Petermann J, Xu Y, Loos J, Yang D (1992) Polymer 33:1096
29. Lotz B, Straupe C, Schumacher M, Wittmann JC, Lovinger AJ (to be published)
30. Smith P, Lemstra PJ, Pipers JPL, Kiel AM (1981) Colloid Polym Sci 259:1070
31. Takeno A, Okui N, Kitoh T, Muraoka M, Umemoto S, Sakai T (1991) Thin Solid Films 202:205
32. Wicker A (1990) thèse Université de Grenoble

33. Kawaguchi A, Okihara T, Murakami S, Ohara M, Katayama K, Petermann J (1991) J Polym Sci Part B Polym Phys 29:685
34. Okihama T, Kawaguchi A, Ohara M, Katayama K (1990) J Cryst Growth 106:33
35. Makinson KR, Tabor D (1964) Proc Roy Soc Lond A281:49
36. Pooley CM, Tabor D (1972) Proc Roy Soc Lond A329:251
37. Bush RF, Haller I, Huggins HA (1974) IBM Techn Discl Bull 16:3077
38. Hansma H, Motamedi F, Smith P, Hansma P, Wittmann JC (1992) Polymer Commun 33:647
39. Folda T, Hoffmann H, Chanzy H, Smith P (1988) Nature 333:55
40. Gorham WF (1966) J Polym Sci A1 4:3030
41. Shen Y, Lotz B, Wittmann JC (to be published)
42. Ihn KJ, Smith P (to be published)

Authors' address:

Dr. J. C. Wittmann
Institut Charles Sadron
6 rue Boussingault
F-67083 Strasbourg, France

# Structural basis of high-strength high-modulus polymers

V. A. Marikhin and L. P. Myasnikova

Ioffe Physico-Technical Institute, St. Petersburg, Russia

*Abstract:* Various ways of producing high-strength and high-modulus flexible-chain polymers are compared. Multistage zone drawing is considered as a promising technique, whose potentialities are still great. It is shown that the mechanical properties of end products drastically depend on many structural parameters of starting material, such as the molecular weight and molecular weight distribution, the morphology, the length and regularity of the folds, the number and type of defects, as well as on the temperature-rate conditions of the drawing. It is emphasized that the transformation of the initial structure into a microfibrillar one on necking occurs through the unfolding of macromolecules and subsequent stress-induced crystallization. With account of the kinetic theory of strength and of the influence of initial morphology on the strengthening and mechanical destruction that compete during the drawing, one can choose an appropriate morphology of the starting material and an optimum drawing regime. The fine structure of ultradrawn polymers is investigated in detail. The structural basis for the discrepancy in experimentally achieved mechanical properties and theoretical estimates are discussed and further improvement of polymer mechanical behavior is considered.

*Key words:* Extreme mechanical properties — structure/property relationship — multistage zone drawing

## 1. Introduction

There has been, formerly and currently, considerable interest in the preparation of extremely strong and stiff polymeric materials, since the vast majority of commercially manufactured fibers and films have stiffness and strengths well below those theoretically estimated for individual molecules. Over the past decade [1—8], promising results have been attained by several research groups in the world, including the Ioffe Institute. Ultrahigh modulus and ultrahigh strength HDPE samples have been obtained (Table 1) by various strengthening techniques with the values of strength and modulus first comparable with theoretical estimates. The enhanced properties, however, have been achieved only for polymers in the form of thin filaments or fibers, while the mechanical characteristics of bulk material are still much lower. This resembles the situation with low-molecular solids, when extremely high strengths close to theoretical ones were obtained for specific aniosmetric samples (crystal "whiskers" and quartz threads [9—11]), but not for bulk samples with a lower strength.

Since ultrahigh mechanical properties have been attained only for UHMWPE, it is very important to understand the singularities of the polyethylene molecular and supermolecular structure that provide material with extraordinary characteristics and then to apply this understanding for producing other polymeric materials with enhanced mechanical properties.

## 2. Theoretical estimates and experimental data

The following questions arise from the analysis of the data in Table 1. i) What is the correlation between the achieved values of tensile strength $\sigma$, elastic modulus $E$ and theoretical estimates? ii) Is there a relationship between the morphology and mechanical properties of oriented polymers? iii) Which of the available techniques is the most

Table 1. Mechanical properties of oriented PE samples produced in various ways

| | | |
|---|---|---|
| Drawing of gel-crystallized UHMWPE | 6.0—7.0 | 150—200 |
| Drawing of single crystal mats of UHMWPE | 2.0—5.0 | 100—200 |
| Extrusion of single crystal mats of UHMWPE with subsequent drawing | 6.0 | 220 |
| Surface growth technique for UHMWPE | 3.0 | 20—100 |
| Surface growth technique with subsequent drawing of UHMWPE | 4.0—6.0<br>10% ⩽10.0 | 120-130 |
| Drawing of reactor UHMWPE powder | 3.0—3.5 | 130 |
| Solid state extrusion of HDPE | 0.4—0.6 | 70 |
| Multistage drawing of melt-crystallized HDPE | 1.5—2.0 | 100 |
| Conventional commercial HDPE samples | 0.15—0.4 | 5 |

perspective for preparation of polymers with enhanced mechanical properties? It is evident that the upper limits are the magnitudes of $\sigma$ and E of idealized chains or perfect crystals. They have been found primarily for PE from molecular mechanics [12—14], the dynamic theory of crystalline lattices [15—16], and quantum-mechanical claculations [17—19]. Of importance is the choice of the pair interaction potential, especially in case of large deviations of bond legnths, valence and rotation angles from their equilibrium values, combined with a good understanding of the polymer electron structure. The estimates of the longitudinal modulus E for PE trans-chains for various interaction potentials differ by a factor of 3 (from 180 GPa to 490 GPa). Typically, E of 200—250 GPa is used [20], which is consistent with many x-ray data [20, 21]. However, the calculations performed for PE with the electron structure close to that derived from recent neutron scattering experiments [22] give E equal to 300—350 GPa, which are probably the most realistic figures.

Tensile strength of PE is estimated to be $\sigma \simeq 0.1\, E$, i.e., from 18 GPa to 49 GPa (in accordance with the accepted value of elastic modulus). For a long time, the discrepancy between the experimental and theoretical data has been of no practical use, because the experimental values were 2—3 orders of magnitude lower. However, the situation has changed, and it is important to know today what values are ultimately achievable and what are the perspective for further increase in strength and modulus. One should note that for the majority of low-molecular ''whiskers'' or etched quartz filaments, the theoretical and ultimate experimental strengths differ by approximately a factor of two, and this difference still remains to be explained [9, 23]. One of the reasons for this discrepancy may be the thermal motion of atoms and thermal fluctuations in solids, which have not been yet taken into account. The theoretical strength is usually calculated for 0, K. it has been shown in terms of the kinetic theory of fracture [24] that the tensile strength of polymers at room temperatures is by 20—65% lower than at $T \rightarrow 0$,K [25]. It is lower by 25% for PE which means that the ultimately achieved strength at room temperature can not be higher than $\sigma$ = 15—35 GPa for this polymer. One can see from Table 1 that the highest experimental data are much lower, so we should not be under a delusion of the results achieved. Thus, there is a large potentiality for enhancing mechanical properties even for PE. The most plausible explanation of the reduced strength is the heterogeneous structure of oriented polymers.

## 3. Approaches to the production of strong and stiff flexible-chain polymeric materials

It is apparent that to obtain high-strength and high-modulus materials, a parallel alignment of macromolecules is needed. The latter may be achieved in two ways. First, the molecules can be aligned through the transition from a coil to an extended macromolecule that occurs in the gradient shear field during the flow of polymer solution and melt (Fig. 1a). The extended state of macromolecules is fixed during subsequent crystallization in the form of extended chain fibrillar crystals (ECC). This phenomenon is the basis for stress-induced crystallization technique developed by Pennings [26], Porter [1, 27], Baranov-Elyashevich-Frenkel [28, 29] and carried out in one stage.

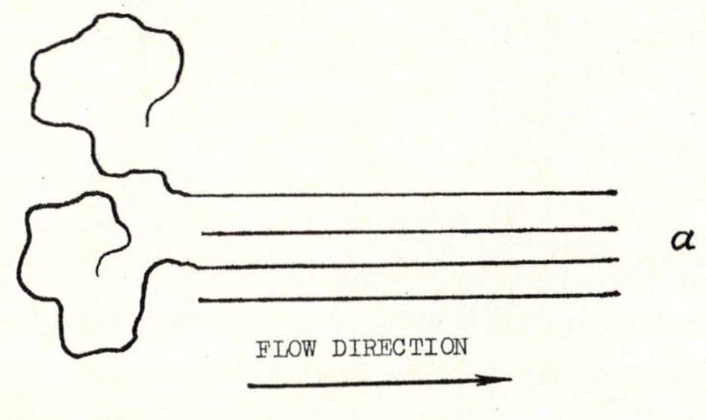


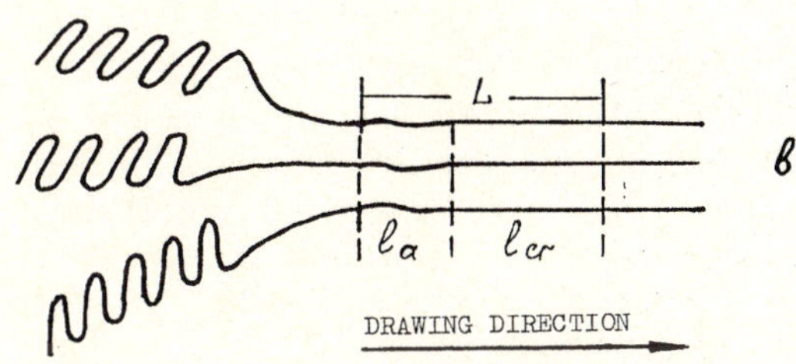


Fig. 1. The model of molecular orientation during a) stress-induced crystallization and b) drawing. $L$ is a long period, $l_a$ is a disordered region, $l_{cr}$ is a crystalline part

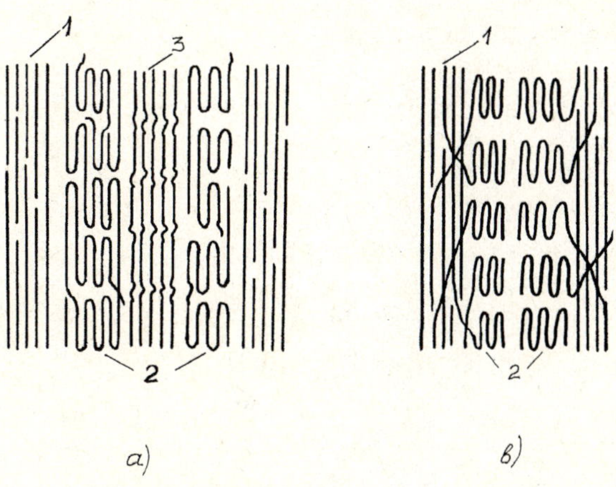


Fig. 2. Structural models of polymers produced through a) solid-state extrusion, b) flow-induced crystallization. Numbers 1, 2, and 3 indicate ECC fibrils, Peterlin-type and Hess-type fibrils, respectively

Second, this can be done through the solid phase transition from a lamellar folded crystal to a fibrillar crystal arising from unfolded chains oriented along the tensile stress (Fig. 1b). This is materialized in the conventional technique of deformation strengthening (drawing) normally performed in several stages [1, 27, 30].

The mechanical characteristics of samples prepared by the technologies based on the above approaches differ greatly (Table 1), which may be accounted for by their specific supermolecular structure. In spite of the attractiveness of the onestage technique of stress-induced crystallization, it is not likely to yield polymers with high values of $\sigma$ or $E$ because of the formation of a composite supermolecular structure with a nonuniform stress distribution over the sample cross section. This type of samples (Fig. 2) can be regarded as consisting of ECC fibrils surrounded by conventional fibrils with a long-period structure, described by the Hess-Hearle or/and Peterlin-Hosemann models [26, 31] (in the case of solid state extrusion) or as having shish-kebab morphology (in the case of flow-induced crystallization). The ECC frame provides high rigidty, but because of the large number of defects (primarily, folds) the ultimate strength, comparable with the theoretical estimates, cannot be achieved for these samples unless they are subjected to additional orientation. This is possible in some cases [4, 32] but is more often unfeasible. Solid state drawing, on the other hand, produces identical microfibrils providing a more uniform stress distribution over the cross-section of the loaded samples. This makes this method promising by itself or in combination with the first one. It is important to study the specific features of the microfibrillar structure and to understand how we can affect it physically and chemically to further enhance its mechanical properties.

## 4. The morphology of drawn polymers

### *4.1. Rearrangement of an unoriented into a fibrillar structure*

Numerous structural data show [30, 33—35] that there are two levels of heterogeneity of a drawn polymer. The polymer bulk consists of macro- and microfibrils (Fig. 3). There are several hundred macromolecules in the cross section of a microfibril. In turn, the cross-section of a macrofibril contains several hundred microfibrils. Both micro- and macrofibrils are interconnected by tie molecules. Analysis of the orientation process of in terms of an individual chemical bond or an individual macromolecule seems to be inadequate, if the real

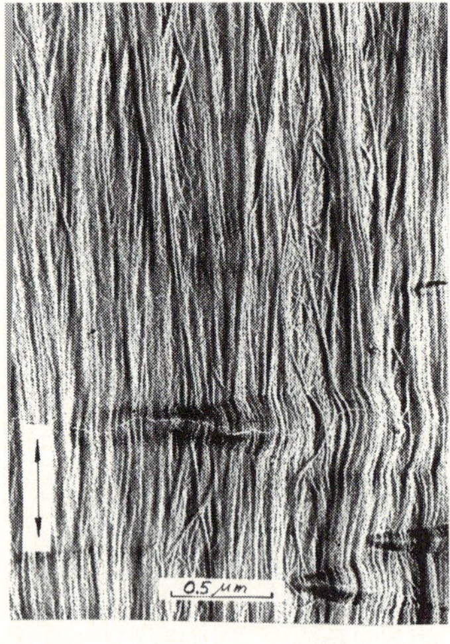


a

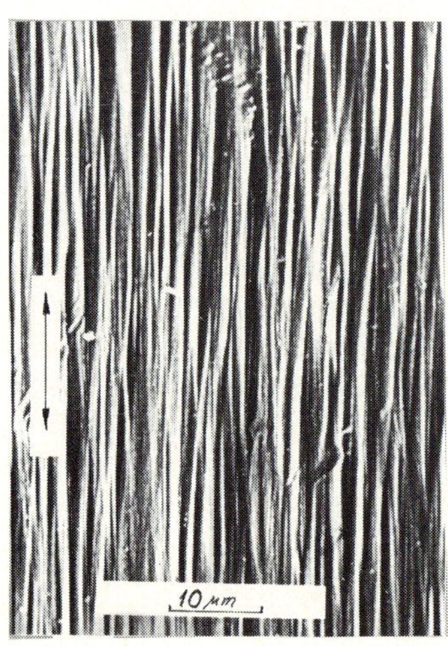


b

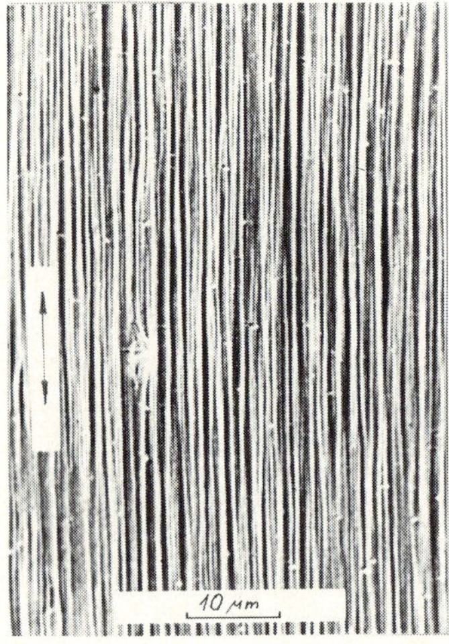


c

Fig. 3. The electron micrographs of microfibrils in PE drawn up to $\lambda$ = 20 (a), replica, transmission microscope, and microfibrils in PE drawn up to $\lambda$ = 10 (b) and $\lambda$ = 30 (c), scanning microscope. Arrows indicate the drawing direction

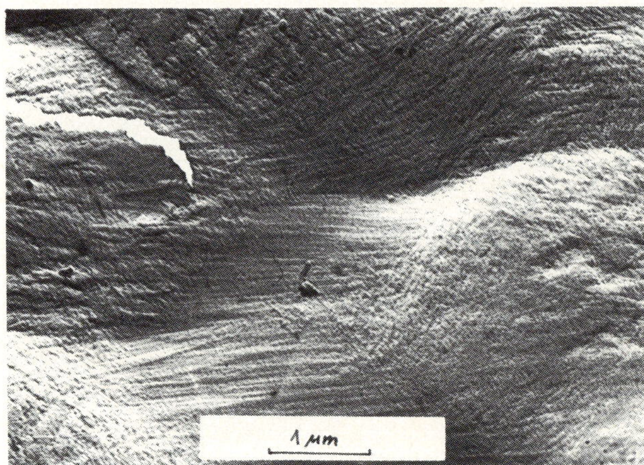


Fig. 4. The replica from the surface of nylon-6 film stretched by 20% at room temperature

supermolecular structure of the oriented polymer is ignored. It should be emphasized that short oriented fibrils are already formed in micronecks at the early stages of macroscopic stretching (Fig. 4) of a few dozen percent [30, 36, 37]. As this takes place, the total elongation of the sample can be hundreds or even tens of thousands percent: for instance, melt-crystallized HDPE samples can be drawn up to the draw ratio $\lambda$ = 30—40 and gel-crystallized UHMWPE samples as well as UHMWPE single crystal mats at $\lambda$ = 200—800. The analysis of our electron microscopic and x-ray data together with the published findings have enabled us to conclude that the initial stage of drawing occurs through the unfolding of macromolecules with subsequent recrystallization and the formation of microfibrils [30, 36—42]. This process is often accompanied by necking. Further drawing is provided by plastic deformation of the microfibrillar structure. Since the samples of one and the same polymer may have identical transverse sizes of their micro- and macrofibrils but different mechanical properties [30, 43, 44], it is natural to suppose that the properties of oriented polymers are primarily controlled by defects localized inside microfibrils. Besides, disordered intermicrofibrillar regions may also play a distinct role. Note that in oriented samples obtained by any technique the internal heterogeneity of microfibrils is generated at the initial stages of orientation, on necking (Fig. 5), which is clear from small-angle x-ray scattering data [30, 45, 46]. The heterogeneity represents a periodic alternation of ordered (crystalline) and disordered (amorphous) regions with different densities and a long period $L$ of several tens or hundreds Å.

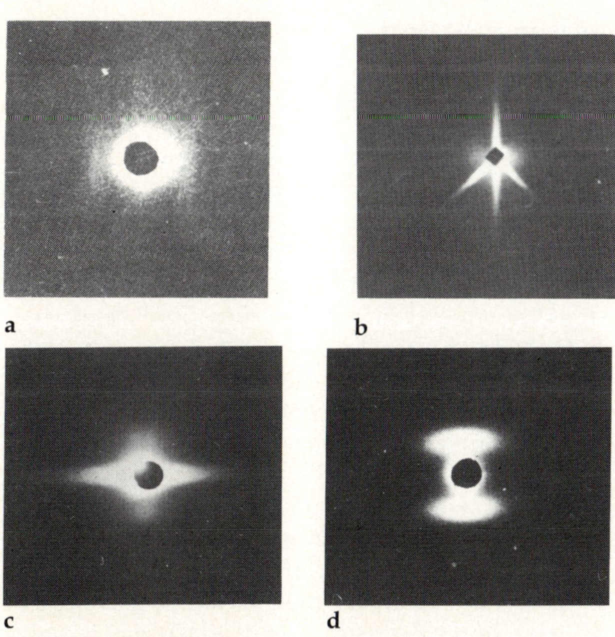


Fig. 5. SAXS patterns from PE samples at the initial stage of orientation through: solid-state extrusion (a), surface growth technique (b), drawing of gel-crystallized (c) and melt-crystallized (d) samples. Numbers 1 and 2 on (b) indicate the instrumental scattering from the crossed Kratky collimators

At present, there are different notions of the structure of microfibrillar amorphous regions (see Fig. 6 and refs. [30, 35, 47]). Judging by the experimental data, the Hess-Hearle model is correct [30]. This means that the majority of chains in the disordered intrafibrillar regions are tie chains. Earlier [48, 49], we suggested a model of the structure of amorphous regions as consisting of a set of macromolecular segments with different lengths

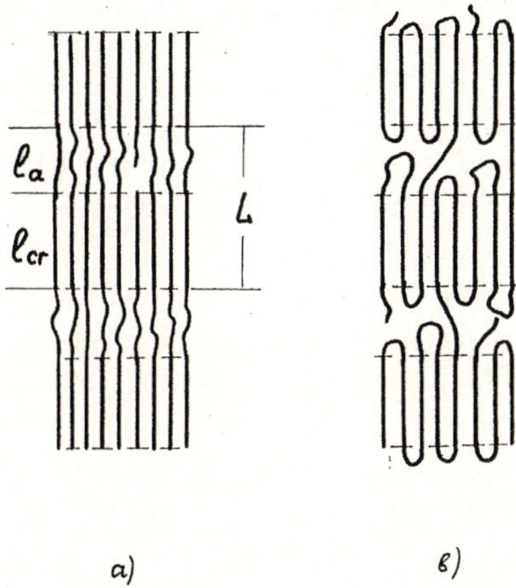


Fig. 6. The Hess-Hearle (a) and the Peterlin (b) models of microfibrillar structure. The notation is the same as in Fig. 1

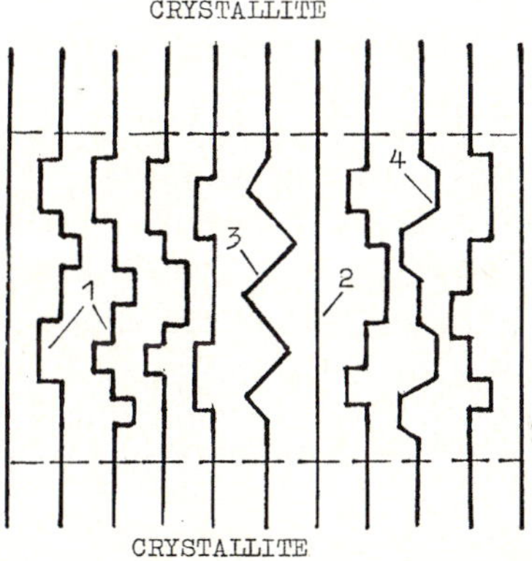


Fig. 7. The model of microfibrillar disordered region proposed by Marikhin [48]. The tie segments is formed by crankshaft-type conformers (1), trans-conformers (2), GG-conformers (3) and TGT-conformers (4)

and a conformation of the crankschaft type (Fig. 7). The major conformational defects here are 2G1 defects, in the classification of Pechold [50]. Such defects can be found also in crystallites but at lower concentrations. The tie molecules become equal in size to one another during the drawing due to the migration of kinks through the crystallites under the action of tensile or shear stresses. The kinks of opposite signs annihilate. As long as the samples have a pronounced long-period structure, the real properties of polymers are determined primarily by the disordered intrafibrillar regions. So the specific features of the crystallite structure do not play a significant role at this stage of drawing. However, as the structure of amorphous regions approaches that of the crystal at high and ultrahigh orientation ($\lambda >$ 20—30), the specificity of macromolecular packing in crystals begins to manifest itself to a large extent.

### 4.2. *The length distribution of molecular segments in disordered regions as revealed by NMR*

Important information on the structure of disordered regions has been derived from WAXS, SAXS, IR, NMR, DSC and other studies [30, 31, 37, 40, 41, 43, 49, 51—60]. At the initial stages of drawing, when a small-angle x-ray maximum was still observable in the samples and the macromolecules in disordered interlayers exhibited micro-brownian mobility, we found, using broad-line NMR, the length distribution functions of the segments in the disordered regions (see Fig. 8 and refs. [58, 60]). With increasing draw ratio from the neck ($\lambda$ = 7) up to the limit for this starting material ($\lambda$ = 28), the number of short segments grows, while the number of long segments decreases. This means that the tie segment lengths equalize. It is evident that in case of broad length distribution the sample strength is determined mainly by the number of the shortest segments. From these data, the fraction of such segments has been calculated for commercially drawn HDPE (Fig. 9). One can conclude that

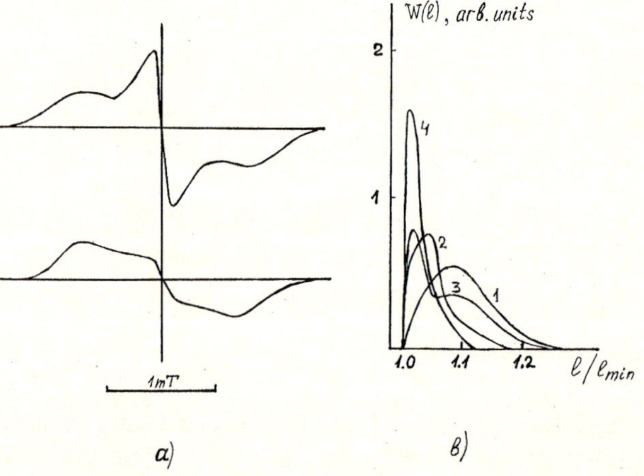


Fig. 8. NMR spectra (a) and the tie molecule length distribution function W(1) for the drawn samples of varying draw ratio: $\lambda$ = 7 (curve 1), $\lambda$ = 16 (curve 2), $\lambda$ = 20 (curve 3), $\lambda$ = 30 (curve 4)

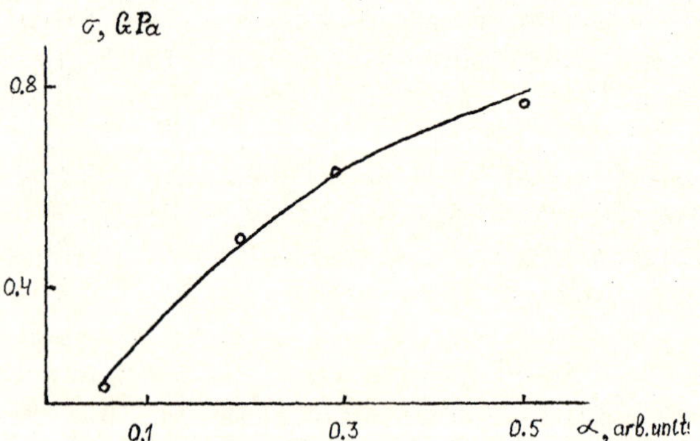


Fig. 9. The dependence of tensile strength of HDPE samples on the fraction of the shortest segments

i) the fraction of load-carrying chains is high (not less than 50%) even for the samples with relatively low strength ($\sigma \simeq$ 0.8 GPa), which corresponds to the fibrillar model with a large number of tie molecules; ii) at high draw ratio $\lambda$, the difference in the tie segment lengths becomes small; iii) since these segments still make contribution to the narrow component, their length is much longer (by 20—30%) than the distance between the crystallites in microfibrils that act like microclamps.

It is known [14] that the introduction of only one kind of defect in a transchain reduces the chain modulus of elasticity by almost a factor of two, so poorly oriented samples with a high concentration of kinks cannot have high mechanical characteristics. If we further increase the draw ratio or perform the drawing more efficiently, the narrow component often disappears in the NMR spectrum (Fig. 8a), so it becomes difficult to find the length distribution of tie molecules. However, this type of spectrum indicates that the macromolecular segments become more extended and uniform, i.e., both intra- and interfibrillar macromolecules with a high degree of cooling disappear and the number of defects in disordered regions decreases appreciably. Obviously, the difference in lengths for the tie chains is units of Å for a highly oriented polymer. Is that small or large?

The deformation of C—C bonds ($\Delta_r$) under load ($f$) can be described by the equation

$$\Delta r = \frac{1}{b} \ln \left( \frac{2}{1 + \sqrt{1 - f/F_m}} \right),$$

where $b$ is a constant and $F_m$ is a limit of the C—C bond strength [49]. Naturally, the difference in length between the chain segments interconnecting the neighboring crystallites results in the difference in the load applied to them, when the polymer is extended.

It is easy to show with this equation that a difference only in several C—C bonds leads to a drastic reduction in the load experienced by a given chain segment oriented along the neighboring fully extended chains. These estimates show to what small value the difference in lengths should be reduced in the disordered region during the drawing as compared with the initial one in order to obtain high strengths and moduli.

### 4.3. *Overstressed tie molecules*

For evaluation of the real number of load-bearing tie molecules, the effect of displacement of the IR absorption bands under load (Fig. 10) can be used. There are methods that can define stresses, acting on the coformers of different structure, from the displacement magnitude [61]. These studies, in combination with structural methods [54], have shown that in typical flexible-chain polymers (PP, PET, nylon-6, etc.), with relatively low strength $\sigma <$ 1 GPa, there is only about 3—10% of the total number of tie chains that can bear loads by an order of magnitude greater than the average microscopic stress. The IR data indicate that with increasing $\lambda$ the concentration of irregular segments lowers and the structure of disordered regions becomes more homogeneous. This reduces the overstresses on the tie molecules [54] since an increasingly larger number of tie segments takes on the external load.

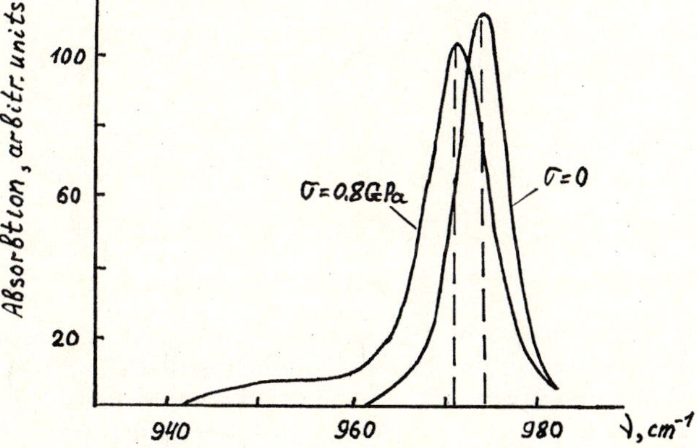


Fig. 10. The shift of the IR absorption band 975 cm$^{-1}$ from loaded PP samples

### 4.4. *Crystalline bridges, tie trans-chains or coherent crystallites?*

That the structure of a disordered region becomes more perfect is evidenced by the WAXS data on the sizes of coherent scattering regions. This is frequently, but not always correctly, attributed to the sizes of crystallites. Typically, the longitudinal dimensions of crystallites $l_{002}$ increase with increasing draw ratio [1, 30, 62, 63]. The most interesting fact observed by a number of authors is that the size of the coherent scattering regions may be larger than the long period [41, 62—64]. This effect was observed for highly oriented PE obtained from both the melt- and the gel-crystallized samples. The ordinary interpretation implies the formation of large crystals, $l_{002} > L$. Since in the frame of a microfibrillar model [44, 45] it is impossible to explain the formation of a crystallite with the longitudinal sizes larger than $L$ and the transverse size equal to that of a microfibril, one should assume that all molecules in the disordered regions are tie molecules. Then the crystallites can grow at the expense of additional crystallization of the extended macromolecular portions in the disordered intrafibrillar regions. In this case an important question is whether the interfaces between the crystals and the amorphous regions can entirely disappear, so that a set of crystallites with any size $l_{002}$ multiple to $L$ can be formed, up to fibrillar crystals of the ECC type.

What do we know about this problem?

It follows from SAXS that at high $\lambda$ the densities of two regions of the long period become almost equal to each other, $\Delta\rho \rightarrow 0$, i.e., this method, beginning from some $\lambda$ well below ultimate $\lambda$, is not sensitive to the boundaries between two regions, so that the whole microfibril can be considered as a needle crystal. As evidenced by WAXS, this is not the case, however, even when we formally accept $l_{002} > L$, a distribution curve for crystallite sizes can be derived from the WAXS data. A method for calculation of the size distribution of these apparent crystallites was suggested by Zubov et al. [65]. Figure 11 shows a large number of crystallites with sizes $l_{002} < L$ and with $l_{002} > L$, up to $l_{002}^{\text{eff}} \sim 5L$, for apparent crystallites. It is clear from these formal calculations that even in ultaoriented samples obtained from mats of single crystals or gel-crystallized samples, the interfaces between the crystal and amorphous regions cannot be eliminated for the majority of long periods. This is quite natural,

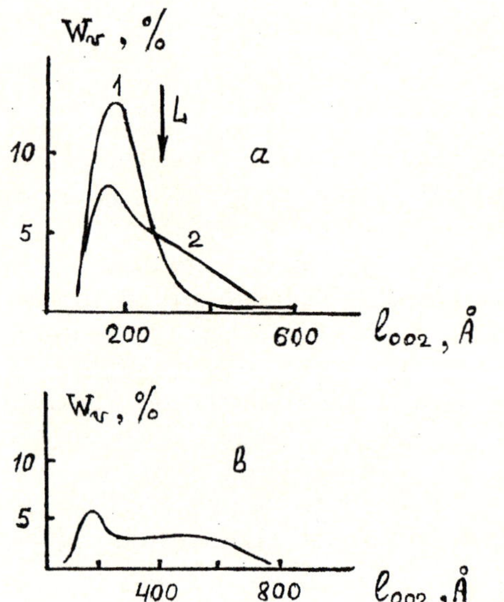


Fig. 11. A typical crystallite size distribution curve for a): melt-crystallized drawn HDPE with $\lambda$ = 5 (curve 1) and 30 (curve 2); the arrow indicates the long period $L$, and for b): drawn single crystal mat of UHMWPE with $\lambda$ = 400 [60, 65]

because WAXS is very sensitive to small distortions of the order, by a few fractions of an Å, in a coherent arrangement of atoms. Therefore, conformational defects of a definite type must accumulate at these interfaces, as they cannot diffuse further through the crystal.

Do these coherent scattering regions with the effective size $l_{002}^{eff} > L$ really exist? Initially, a number of authors, and we among them [1, 47, 63], supposed that large crystals, ofen called crystalline bridges, are formed. Later theoretical analysis showed, however, that this effect can be attributed to purely diffraction phenomena associated either with the contribution from one-dimensional diffraction [66] or with the formation of coherent arrangement of neighboring crystallites (with the sizes lesser than a long period) in a microfibril [67, 68]. One-dimensional diffraction is considerable if there are long portions of individual extended chains that are randomly distributed over the fibril cross-section and pass from one crystallite into another through a disordered region or through several neighboring long periods. A large coherent scattering region can arise also from neighboring small crystallites that, in case of a small difference in lengths of the chain segments in the disordered region, can be arranged in such a way that they scatter in phase with each other. This leads to narrower [002] peak and its higher intensity which was erroneously interpreted as an increase in the crystallite size. Since the x-ray diffraction intensity is proportional to the number of scattering atoms, we believe that the coherent crystallite arrangement makes a greater contributions to the scattering, more so than one-dimensional diffraction. Therefore, it is unlikely that crystallites with a size $l_{002} > L$ appear during the drawing. These data also point to a significant perfection of the microfibrillar structure and equalization of the tie segment lengths in disordered regions by as much as a few units or fractions of an angstrom. As a result, the structure of an initially strongly heterogeneous microfibril approaches that of a needle-like defect crystal.

### *4.5. Thermal properties of oriented polymers*

In addition to the improvement of mechanical characteristics, structural perfection of oriented polymers should lead to better thermal properties — to higher melting temperature and thermal stability and smaller thermal shrinkage [1, 30, 47].

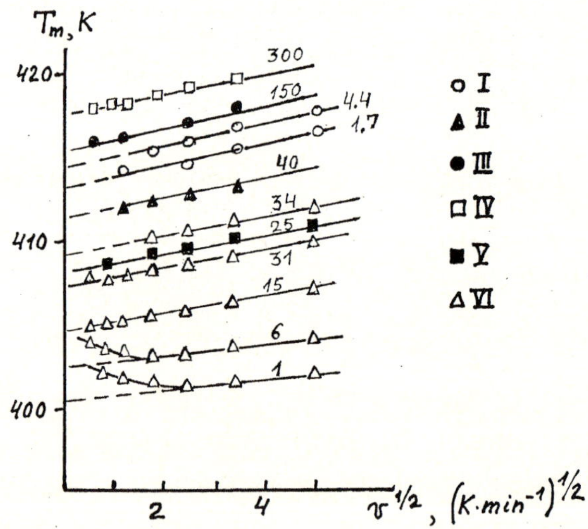


Fig. 12. The dependence of melting temperature on the heating speed of PE samples produced through multistage zone drawing of solution-crystallized polymer (○), gel-crystallized (▲) and fibers (●), melt-crystallized quenched films (△) and through solid-state extrusion of melt-crystallized polymer (■) and single crystal mats subjected to conventional drawing following the extrusion (□). The figures near the curves indicate $\lambda$

Moreover, the study of melting of oriented samples can give additional information on the state of macromolecules in disordered regions, because the melting characteristics of crystallites can be appreciably affected by the number and conformational structure of the tie molecules [56, 69, 70]. Figure 12 shows the dependence of the melting temperature $T_m$ on the heating rate $V$ that allows real $T^r_m$ and real melting interval $\Delta T^r_m$ to be determined (Table 2). It is known that the process of polymer melting is characterized by a broad phase transition, $\Delta T_m \sim 10$ K, and by a lower current melting temperature as compared with the equilibrium: $T_m < T^0_m = 416$ K [69]. The data show that as the degree of orientation grows, the dependence $T_m(V^{1/2})$ shifts towards higher temperatures, and $T_m$ increases from $T_m = 402.7$ K ($\lambda = 6$, neck, melt-crystallized sample) to $T_m = 416$ K ($\lambda = 200$, gel-technique). The melting peak narrows sharply and its height increases. These data point to gradual disappearance of the interfaces between a crystal and a disordered region and to better perfection of the microfibrillar structure. The estimates of the end surface energy $\gamma_i$ obtained from the Thomson-Gibbs equation [69] also confirm these facts:

$$\lambda = 6 \qquad \gamma_i = 100 \text{ erg} \times \text{cm}^{-2}$$

$$\lambda = 34 \qquad \gamma_i = 20 \text{ erg} \times \text{cm}^{-2}$$

$$\lambda = 200 \qquad \gamma_i = 0 .$$

It is clear from the DSC data that the microfibril approaches an ideal crystal. With the Flory relation [71], we can find the intrachain cooperativity parameter $\nu$ for the order-disorder transition (in our case, crystal-melt) from the halfwidth of the melting peak

$$\nu = 2R(T^r_m)^2/\Delta T^r_m \cdot \Delta H .$$

For our samples with $\lambda = 34$, $\nu = 30$ nm and with $\lambda = 200$, $\nu = 2000$ nm, which means that the sizes of extended chain segments estimated by DSC are several orders of magnitude larger than the those of coherent scattering regions determined by WAXS. In addition, the low values of the melting interval $\Delta T_m = 0.04$—$0.4$ K also point to the absence of appreciable dispersion of the cooperativity paramter. Since $\nu$ is comparable in the order of magnitude with the length of a macromolecule, we can conclude from the DSC data that in highly oriented samples (prepared by the melt and other techniques) the molecules are strongly extended, each passing through hundreds and thousands of crystallites and disordered regions (long periods). This means that a disturbance in the three-dimensional order in intercrystalline regions must indeed be limited only by defects of molecular packing of the disclination type with rotation of the trans-zigzag plane. These were shown by Reneker [72, 73] to be unable to overcome the crystal-amorphous region interface. Thus, the thermodynamic parameter $\nu$ characterizes the general orientation of the majority of chains in microfibrils, so one can expect a correlation between $\nu$ and the sample strength. This indeed is so (Fig. 13) and a direct relationship between mechanical properties and thermal parameters has first been demonstrated.

Table 2. Thermal characteristics of oriented PE samples produced in various ways

| Preparation technology | $M_w$ | $\lambda$ | $T^{true}_m$, K | $\Delta T^{true}_m$, K |
|---|---|---|---|---|
| Surface growth technique with subsequent multistage zone drawing (MZD) | $1.5 \cdot 10^6$ | 2.5 | 414.0 | 0.40—0.50 |
| MZD of gel-crystallized films | $3.5 \cdot 10^6$ | 200 | 414.0 | 0.05—0.07 |
| MZD of gel-crystallized fibers | $1.5 \cdot 10^6$ | 150 | 415.0 | 0.04—0.05 |
| Solid-state extrusion of single crystal mats | $4.5 \cdot 10^6$ | 300 | 417.5 | 0.05—0.07 |
| Solid state extrusion of melt-crystallized billets | $10^6$ | 25 | 408.0 | 1.8 |
| MZD of melt-crystallized films | $10^5$ | 30 | 409.0 | 0.3—0.4 |

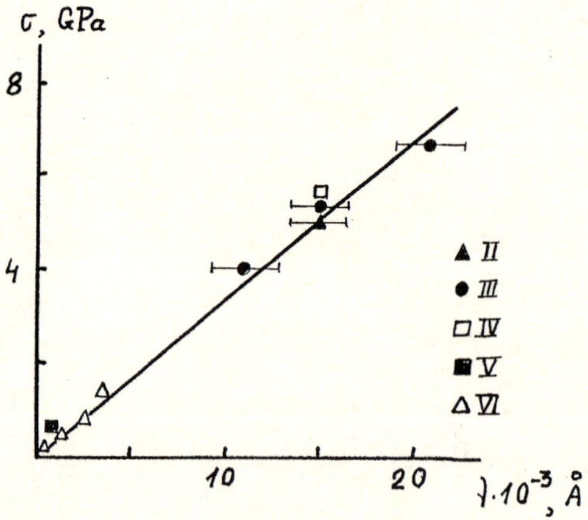


Fig. 13. The dependence of tensile strength of oriented and ultraoriented PE samples on the parameter intrachain cooperativity of melting $\nu$. the notations indicating the techniques used for orientation are the same as in Fig. 12

It is well known that the structural perfection and higher melting temperature lead to a sharp reduction in the thermal shrinkage of highly oriented samples that does not exceed a few percent at temperatures only several degrees away from the melting temperature.

### 4.6. *Possible improvement of polymer mechanical behavior*

Clearly, the properties of oriented polymers are determined primarily by the structure of the most imperfect (defect-containing) intrafibrillar regions. The major reason for the number of defects of these regions is the difference in the lengths of macromolecules connecting neighboring crystallites, which varies with the concentration of the simplest conformational and disclination defects. At present, it is very important to obtain new quantitative information on the defects of this type. Earlier, the problem was to eliminate pronounced disordered intrafibrillar regions, but the major problem today is the elimination of much simpler defects in fibrillar crystals. Unfortunately, they are quite stable. Modern physical methods allow, in principle, detailed studies of defect structure at this level, but such studies are still very few. We hope that in the near future the concepts of defect theory well-elaborated for ordinary low-molecular weight crystals will be successfully applied to the physics of polymers.

It follows from the data presented in this lecture that of primary importance is the microfibril nucleation at the initial stages of deformation, when a long-period structure with highly imperfect disordered intrafibrillar regions is formed. The number of defects of these regions is strongly affected by specific features of the folded surface structure in initial lamellae and the connectedness of the lamellae. It is a priori clear that the more regular is the fold surface, the lesser defects are found in the neck fibrils, i.e., the difference in the chain lengths in the disordered regions is smaller. From this point of view, the best microfibrils could be obtained by drawing individual single crystals. However, this suggestion cannot be proved by direct experiment since it is impossible to measure the mechanical characteristics of extremely thin microfibrillar units arising during single crystal deformation. The next step to an ideal starting material is a single-crystal mat, from which oriented samples with fairly high mechanical properties are obtained. The achieved strength, however, is still far from the theoretical estimates [8]. The optimization of the drawing regime should probably improve their mechanical characteristics. These studies are primarily of academic interest. As far as applied problems are concerned, the most suitable structure for processing is that of gel-crystallized high molecular weight polymers [3]. Scanning microscopy and SAXS data show that the stacks of lamellae are formed in gel-crystallized samples. Our data indicate the significant differences of these lamellae from the lamellae formed during melt crystallization, namely, they have a low thickness $L \simeq 7$ nm as compared with ordinary $L \simeq 24$—$26$ nm, a thinner fold surface $l_a \simeq 2$ nm, against $l_a \simeq 5$ nm, and a more regular structure of the fold surface, or a much lower content of irregular conformers.

Melt-crystallized samples have a structure most unsuitable for successful strengthening [74—76]. The fold surface thickness is approximately equal to that of the crystalline core owing to a large number of long irregular loops and numerous tie molecules. The latter often form fibrillar crystals, especially at high $T_{cr}$ [74, 77], and this affects adversely the structural rearrangement during orientation, so that no high-strength and high-modulus material can be obtained [53, 78]. Perfection of fold surfaces in the initial lamellae, for instance by annealing, may improve the mechanical behaviour of oriented polymers. The processing of nascent polymers by

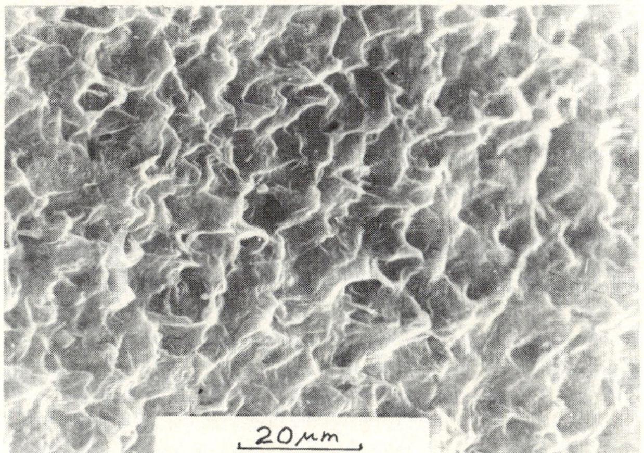


Fig. 14. Scanning micrograph of a gel-crystallized UHMWPE film

a technique similar to that in powder metallurgy also offers promise.

## 5. Conclusion

The drawing in the solid state is a perspective method for producing high-strength and high-modulus flexible-chain polymers since its potentialities have not been exhausted yet. What can promote further progress in this field?

Of great importance are:

i) the right choice of initial molecular characteristics (molecular weight and molecular weight distribution),

ii) the creation of optimum morphology of starting material by varying the crystallization conditions [3, 79, 80, 81],

iii) the optimization of the orientation process by using the structural-kinetic approach to strengthening [30, 40, 51, 53]. One should keep in mind that two phenomena occur simultaneously during the drawing: strengthening due to the orientation of macromolecules and strength losses due to the mechanical and thermal destruction, relaxation, and annealing of oriented polymers. The kinetics of both phenomena varies dramatically not only with the initial morphology but also with the temperature-rate regime of the drawing. Since the kinetics depends exponentially on temperature and stress, the choice of orientation conditions should be made very carefully.

Of special interest is the formation of a neck. The generation of microfibrils with a nematic structure on necking may facilitate the migration of defects through the crystallites and their annihilation, which favors the production of high-strength and high-modulus materials. The longer the nematic state will be preserved during the drawing the better mechanical properties of the end product can be attained. The possibility of attaining a nematic state at the initial stage of multistage zone-drawing has enabled us to produce extraordinary PETP samples with record $\sigma$ = 1.86 GPa and $E$ = 34 GPa. The use of the above principles in producing HDPE samples (conventional $M_w$ about 100000—200000) with enhanced mechanical properties has also yielded promising results ($\sigma$ = 1.8—2.0 GPa and $E$ = 100 GPa).

However, increasing tensile strength and modulus of elasticity of polymers up to ultimate values inevitably give rise to production of one-dimensional defect crystals (paracrystals) which loose the intrinsic polymer properties (flexibility, the large reversible deformation, the resistance to multiple bends, etc.). This results in a specific fracture of ultrastrong and ultrastiff samples because of kink bands arising on their deformation. The latter lowers the durability and limits the application of such materials. They are largely suitable as fillers in composites.

## References


1. Chiferri A, Ward IM (eds) (1979) Ultra-high Modulus Polymers. Applied Science Publishers
2. Smook I, Torts JC, van Hutten PF, Pennings AJ (1980) Polym Bull 2:293
3. Smith P, Lemstra PJ (1979) Makromol Chem 180:2983
4. Marikhin VA, Myasnikova LP, Zenke D, Hirte R, Waigel P (1974) Polym Bull 14:287
5. Savitskii AV, Gorshkova IA, Frolova IL, Shmikk GN (1984) Polym Bull 12:195
6. Kanamoto T, Tsaruta A, Tanaka K, Takeda M, Porter RS (1988) Macromolecules 21:470
7. Tovmasyan Yu, Takahashi K, Kanamoto T (1989) Reports on Progress in Polymer Physics in Japan 32:147—150
8. Konstantinopolskaya MB, Chvalun SN, Selikhova VI, Ozerin AN, Zubov YuA, Bakeev NF (1985) Vysokomol Soedin B27:538—541
9. Kelly A (1969) Strong Solids (Oxford University Press)
10. Berezhkova GV (1969) Nitevidnye krystally (Moscow: Nauka), p 158

11. Bartenev GM (1974) Sverkhprochnye i vysokoprochnye neorganicheskie stekla. Mosco, p 240
12. Treloar RLG (1960) Polymer 1:95
13. He T (1987) Makromol Chem 188:2489
14. Gajdos J, Blecha T (1987) Materials Chemistry and Physics 17:405
15. Kobajashi M (1979) J Chem Phys 70:509
16. Wobser G, Blasenbrey S (1970) Kolloid-Zeitschr 241:985
17. Boudreaux DS (1973) J Polym Sci, Polym Phys ed 11:1285
18. Karpfen A (1981) J Chem Phys 75:283
19. Suhai S (1986) J Chem Phys 84:5071
20. Holliday L (1975) Structure and properties of oriented polymers. Ch 7, Ward IM (ed) (London: Applied Science Publishers)
21. Sakurada I, Ito T, Nakamae K (1966) J Polym Sci C15:75
22. Tashiro K, Kobayashi M, Todokoro H (1978) Macromolecules 11:914
23. Macmillan NH (1972) J Mater Sci 7:239
24. Zhurkov SN (1965) Intern J of Fracture Mech 1:311
25. He T (1986) Polymer 27:253
26. Pennings AJ (1979) Makromol Chem, Suppl 2:99
27. 1983 Development in oriented polymers. Ward IM (ed) (London: Applied Science Publishers), vol 1, vol 2
28. Elyashevich GK, Baranov VG, Frenkel SYa (1979) J Macromol Sci B13:255
29. Elyashevich GK (1982) Advances Polymer Science 43:205
30. Marikhin VA, Myasnikova LP (1977) Nadmolekulyarnaya struktura polimerov (Leningrad: Khimia)
31. Marikhin VA (1985) Abstracts of 17th Europhysics conference on Macromolecular Physics "Morphology of Polymers" L6, (Prague)
32. Smook I, Torts JC, van Hutten PF, Pennings AJ (1980) Polym Bull 2:293
33. 1964 Fibre structure. Hearle JWS, Peters RH (eds) (Manchester and London: The Textile Inst., Butterworths)
34. 1979 Applied Fibre Science. Happey F (ed) (London-N.Y.: Academic Press), vol 1, p 564, vol 2, p 553
35. Peterlin A (1975) Colloid and Polym Sci 253:809
36. Zhurkov SN, Marikhin VA, Myasnikova LP, Slutsker AI (1965) Vysokomol Soedin 7:1041
38. Wignall GD, Wu W (1983) Polym Commun 24:354
39. Adams W, Yang D, Thomas EL (1986) J Mater Sci 21:2239
40. Myasnikova LP (1986) Plaste und Kautschuk 33:121
41. Marikhin VA, Myasnikova LP (1991) Makromol Chem., Macromol Symp 41:209
42. Tuo Min Liu, Juska TD, Harrison IR (1988) Polymer 27:247
43. Marikhin VA, Myasnikova LP (1971) Mech polim 2:364
44. Peterlin A (1975) Int J Fracture 11:761
45. Hay IL, Keller AJ (1966) J Mater Sci 1:41
46. Peterlin A, Balta-Calleja FJ (1970) Kolloid-Z u Z Polym 242:1093
47. 1975 Structure and properties of oriented polymers. Ward IM (ed) (London: Applied Science Publishers), p 500
48. Marikhin VA (1977) Fizika tverd tela 19:1036
49. Marikhin VA (1979) Acta Polymerica 30:507
50. Pechold W (1971) J Polym Sci, Polym Symp C32:123
51. Marikhin VA (1984) Makromol Chem, Suppl 7:147
52. Marikhin VA, Myasnikova LP, Sutchkov VA, Tukhvatullina M, Novak II (1972) J Polym Sci C38:195
53. Marikhin VA, Myasnikova LP (1977) J Polym Sci, Polym Symp 58:97
54. Friedland KJ, Marikhin VA, Myasnikova LP, Vettegren VI (1977) J Polym Sci, Polym Symp 5:185
55. Egorov EA, Zhizhenkov VV, Marikhin VA, Myasnikova LP (1990) J Macromol Sci Phys B29:129
56. Vettegren VA, Egorov VM, Marikhin VA, Myasnikova LP, Sirota AG (1991) Abstracts of Intern Conf "Deformation, Yield and Fracture of Polymers", 47/1—47/4 (Cambridge)
57. Marikhin VA, Valtonen AI, Zolotarev VM, Mirza AV, Myasnikova LP, Chmel AE (1990) Vysokomol Soedin A32:2378
58. Zhizhenkov VV, Egorov EA, Marikhin VA, Myasnikova LP (1985) Mech komposiz materialov 2:354
59. Marikhin VA, Myasnikova LP, Victorova NL (1976) Vysokomol Soedin A18:1302
60. Egorov EA, Zhizhenkov VV, Marikhin VA, Myasnikova LP, Gann LA, Budtov VP (1985) Vysokomol Soedin A27:1637
61. Vettegren VI, Novak II, Friedland KI (1975) Intern, J Fracture 11:789
62. Zubov YuA, Chvalun SN, Selikhova VI, Konstantinopolskaya M, Band Bakeev NF (1988) Zhurnal Fizicheskoi khimii 62:2815
63. Frye CJ, Ward IM, Dobb MG, Johnson DJ (1982) J Polym Sci, Polym Phys ed 20:1677
64. Smith P, Boudet A, Chanzy H (1985) J Mater Sci Lett 4:13
65. Ozerin AN, Zubov YuA (1984) Vysokomol soedin A26:394
66. Chvalun SN, Shiretz VS, Zubov YuA, Bakeev NF (1986) Vysokomol Soedin A28:18'
67. Azriel YeA, Vasiliev VA, Kazarian LG (1989) Vysokomol Soedin A28:810
68. Azriel AYe, Vasiliev VA, Kazarian LG (1989) Vysokomol Soedin A31:2412
69. Wunderlich B (1980) Macromolecular Physics, Crystal Melting (Academic Press, N.Y.), 3:361
70. Berstein VA, Egorov VM, Marikhin VA, Myasnikova LP (1990) Vysokomol Soedin A32:2380
71. Flory P (1961) Proc Royal Sco 49:105
72. Reneker DH, Mazur J (1983) Polymer 24:1387
73. Reneker DH, Mazur J (1988) Polymer 29:3
74. Egorov EA, Zhizhenkov VV, Marikhin VA, Myasnikova LP, Popov A (1983) Vysokomol Soedin A25:693
75. Berstein VA, Egorov VM, Marikhin Va, Myasnikova LP (1985) Vysokomol Soedin A27:771
76. Vettegren VI, Marikhin VA, Myasnikova LP, Popov A, Bodor G (1986) Vysokomol Soedin A28:914

77. Berstein VA, Egor VM, Marikhin VA, Myasnikova LP (1986) Vysokomol Soedin A28:1983
78. Korsukov VE, Marikhin VA, Myasnikova LP, Novak II (1973) J Polym Sci C42:847
79. Gann LA, Marikhin VA, Myasnikova lP, Budtov VP, Myasnikov GD, Ponomareva EL (1987) Vysokomol Soedin A29:1658
80. Gann LA, Marikhin VA, Myasnikova lP, Budtov VP, Myasnikov GD (1988) Vysokomol Soedin A30:573
81. Capaccio G, Ward IM (1974) Polymer 15:233

Authors' address:

Dr. L. P. Myasnikova
Ioffe Physico-Technical Institute
26, Polytekhnicheskaya st.
194021 St. Petersburg, Russia

# Transformations of defect structure of polymers during drawing

N. A. Pertsev

A. F. Ioffe Physico-Technical Institute, Russian Academy of Sciences, St. Petersburg, Russia

*Abstract:* The defects inherent in semicrystalline linear polymers and their behavior during the drawing are considered. These defects are classified in accordance with their relation to different scale levels of polymer morphology and their conformational or crystallographic nature. The topological stability of various structural imperfections is discussed. The transformations of the defect structure of polymers during the drawing are analyzed. Particular attention is given to the annihilation of original defects and the creation of new ones during the rearrangement of the initial supermolecular structure into the microfibrils. The plastic deformation of the fibrillar is considered. A dislocation mechanism is proposed for the slippage of microfibrils during the drawing. Dislocation-disclination models are worked out for the kink bands in oriented polymers. These models permit to calculate the internal stresses induced by kink bands and to analyze the mechanisms of microcracking caused by kinking.

*Key words:* Conformational defects — dislocations — disclinations — kink bands — plastic deformation

## 1. Introduction

It is well known that structural imperfections markedly influence the mechanical properties of solids. For low-molecular crystals the theory of defects is now well-established and has been extensively applied to the modeling of the plastic deformation and fracture of crystals. Theoretical ideas about defects appropriate to polymers are less developed. This is partly due to the fact that, in case of polymers, the problem is complicated by the chain structure of polymer molecules and by the existence of various scale levels of the polymer morphology. Nevertheless, it is worthwhile to study structural imperfections of polymers on the base of the general theory of dislocations and disclinations developed for solids. With this approach we gain the advantage of using for our purposes numerous results obtained earlier for low-molecular crystals.

In this review we first consider modern ideas about the defects inherent in semicrystalline linear polymers. Then, we shall discuss transformation of the defect structure of polymers during the drawing. The analysis performed below is useful for better understanding the structure/property relationships in oriented polymers.

## 2. Classification of defects appropriate to semicrystalline polymers

It is reasonable to classify defects in accordance with their relation to different levels of polymer morphology which correspond to macromolecules, crystallites, lamellae, fibrils, spherulites, and so on. We shall mainly discuss the defects that relate to the scale of transverse sizes of macromolecules and to the level of lamellae and microfibrils. These defects play an important role in the drawing process.

Structural imperfections of molecular scale have dimensions of the order of 10 Å. They are subdivided into conformational and crystallographic defects. The nucleation of defects of the first kind results from conformational transitions occurring in flexible chain polymers [1—3]. The stability of a conformational defect in the absence of external forces is ensured by an intramolecular potential barrier which has to be overcome during the internal rotation around backbone bonds. If we treat the polymer as an elastic continuum, then an elementary defect corresponding to a gauche conformer located between two trans-conformers should be modeled by a pair of wedge and twist disclination loops enveloping the macromolecule (Fig. 1a, b) [4—6]. One should note that disclinations are the line defects of rotation type which here provide bend and twist of chains [4].

When the chain defect is embedded into a crystallite replacing an original regular chain, it is deformed elastically as well as is the surrounding material (Fig. 1c). The disclination model introduced above allows the calculation of such deformations and associated elastic energy of a conformational defect [4, 6]. This is the major advantage of the disclination approach.

The models of more complex conformational defects can now be easily constructed in the form of ensembles of disclination loops. For instance, the widely discussed defect termed "kink" [1, 7—9] is modeled by two pairs of wedge and twist loops with opposite signs (Fig. 2) [6].

The existence and stability of crystallographic molecular defects is ensured by intermolecular forces that bind the chains inside crystallites. Crystallographic defects include dislocation loops and also new disclination loops, which will be termed "external" disclinations in order to distinguish them from "internal" disclination loops of conformational origin [5, 6]. A prismatic dislocation loop enveloping a single macromolecule is formed when a part of it is translated along the chain axis by a distance being equal, for example, to the corresponding lattice period "c" [10, 11]. In turn, the nucleation of an external twist disclination loop results from a rigid rotation of a part of a macromolecule around the chain axis (Fig. 3a, b). The rotated part must have a definite setting angle in the unit cell that corresponds to a new minimum of the interaction energy between macromolecule and the surrounding lattice [5, 6].

An external wedge loop models a source of strains in the form of elastically bent chain which is held in this state by neighboring macromolecules. The widely occurring defect of this type is a 180° loop (Fig. 3c), which is a characteristic element of chain fold surfaces of crystalline lamellae in unoriented polymers [6]. Of course, the purely elastic 180° bend is not energetically favorable. In reality, the chain folding involves conformational transitions so that additional internal wedge loops of opposite sign appear which decrease the elastic energy of a bent chain [4]. Thus, the chain fold should be modeled by a combination of crystallographic and conformational defects.

The appearance of combined defects with these two components is typical of polymer crystals. Such a mixed origin has, for instance, the Reneker defect in polyethylene, which transports $CH_2$-group along the chain. This defect contains a twist of 180°

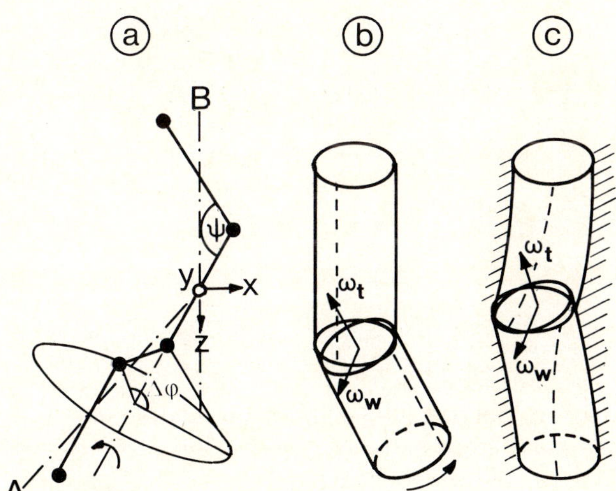


Fig. 1. Geometry of rotation of a skeletal bond in a polymer chain (a) and the formation of internal twist and wedge disclination loops in an isolated chain (b) and in a macromolecule incorporated into a solid polymer (c). Here, $\Delta\varphi$ denotes the variation of internal rotation angle (torsional angle) and $\Psi$ is the valence angle of the chain. $\vec{\omega}_W$ and $\vec{\omega}_t$ are the Frank vectors of wedge and twist loops

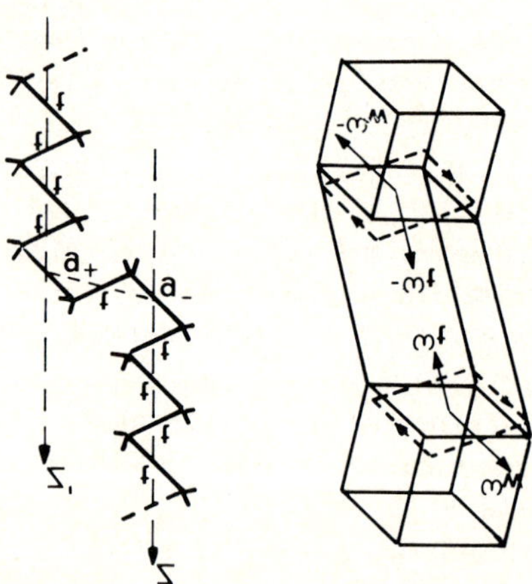

Fig. 2. a) Schematic illustration of a macromolecule containing conformational kink defect. b) Continuum disclination model of a kink in a polymer crystal

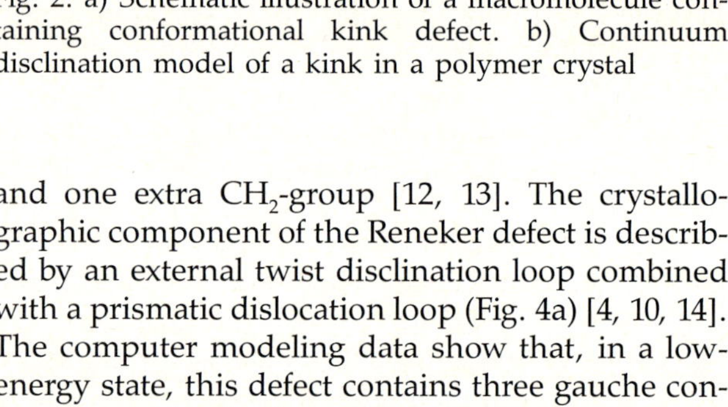

and one extra $CH_2$-group [12, 13]. The crystallographic component of the Reneker defect is described by an external twist disclination loop combined with a prismatic dislocation loop (Fig. 4a) [4, 10, 14]. The computer modeling data show that, in a low-energy state, this defect contains three gauche conformers with altering sign [13]. Thus, the con-

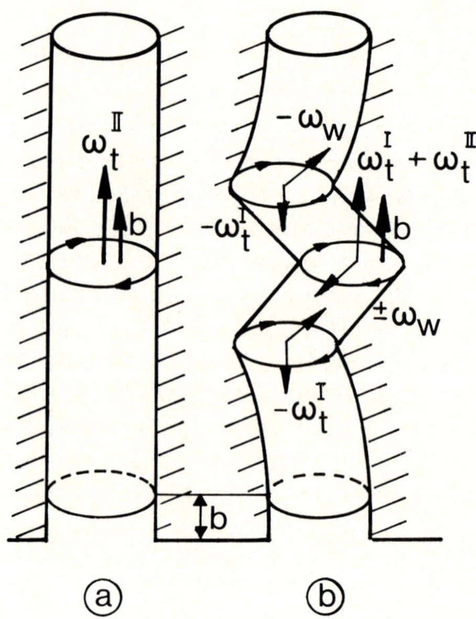


Fig. 4. Dislocation-disclination models of the Reneker defect in polyethylene. a) The combination of the prismatic dislocation loop with Burger vector $\vec{b}$ and the external twist disclination loop with Frank vector $\vec{\omega}_t''$, which describes the crystallographic component of the Reneker defect. b) The complete model including additional internal wedge and twist loops with Frank vectors $\pm\,\vec{\omega}_W$ and $\pm\,\vec{\omega}_t'$

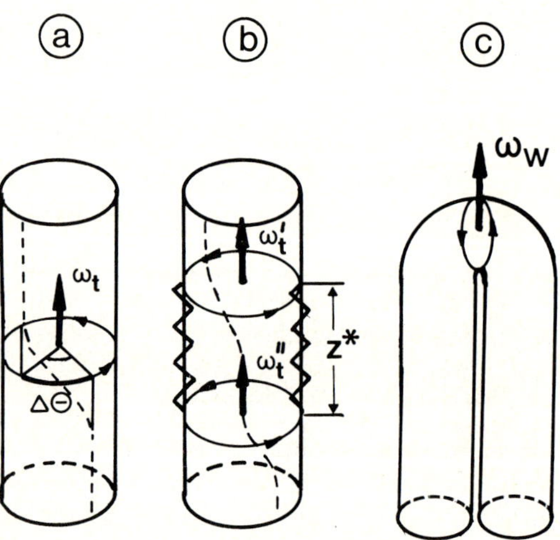


Fig. 3. External disclination loops in a polymer. a) Perfect twist disclination loop of strength $\omega_t = \Delta\Theta$. b) Dissociation of a perfect twist loop into two partial loops with Frank vectors $\vec{\omega}_t'$ and $\vec{\omega}_t''$ [6]. c) Wedge loop of strength $\omega_W = 180°$

tinuum model of a Reneker defect should include also additional internal twist and wedge loops (Fig. 4b).

We conclude the analysis of molecular defects with the consideration of their topological stability. This is necessary for establishing the mechanisms of defect nucleation and elimination. It can be shown that an isolated dislocation or disclination loop which surrounds a single chain or several neighboring chains is topologically stable in a solid polymer. Really, these defects cannot be removed locally without breaking the backbone bonds. They disappear either when a defect reaches a chain end or when it annihilates with a defect of opposite sign. At the same time certain complex defects, such as kinks, for example, should be regarded as topologically unstable. (The kink can be eliminated by means of the local internal rotation of the central skeletal bond in it [8]).

The next scale corresponds to the thickness of lamellar crystallites in unoriented polymers and to the transverse dimension of microfibrils in oriented materials. Defects existing on this level have dimensions of the order of 100 Å. They include, for example, straight edge and screw dislocations with Burgers vectors prescribed by the lattice periodicity [15, 16]. Electron micrographs indicate that in unoriented polyethylene the dislocation density inside lamellae is relatively high (of the order of $10^{11}$ $m^{-2}$) [17, 18]. The chain ends also must be regarded as imperfections of lamellar scale, because the accomodation of a chain end in a crystal lattice results in the formation of paired screw dislocations coming from this defect [19].

Lamellar crystals can also contain two-dimensional imperfections such as stacking faults terminating on partial dislocations and as boundaries of twins and mosaic blocks [3, 17, 20]. However, of most importance is an inherent two-dimensional defect of the chain folded crystallites which is usually termed "fold surface" [3]. It is known that the fine structure of fold surfaces is not very regular. (The surface is characterized by a statistical distribution of fold lengths). Nevertheless, we can state that each fold contains one external 180° wedge loop and approximately the same number of internal wedge loops with opposite sign. This feature will be taken into account in the further analysis.

In oriented polymers the imperfections of microfibrillar scale include disordered layers being three-dimensional defects and the interphase boundaries between these layers and crystallites, which are two-dimensional defects. We expect that specific line defects also should exist here. For example, a shift of a part of microfibril along its axis that is realized by the slippage of this part relative to neighboring microfibrils results in the nucleation of a prismatic quasi-dislocation loop enveloping this microfibril. In turn, the inelastic bend of a disorered interlayer caused by multiple correlated conformational transitions is equivalent to the formation of an effective wedge disclination loop around a microfibril. It will be shown below that such defects may appear during the drawing process.

## 3. Transformations of the defect structure during drawing

In this section the behavior of basic defects during drawing will be considered. We shall discuss only uniaxial extension of polymers having originally a spherulitic morphology at relatively slow rates when plastic deformation is not suppressed. In this situation the deformation process occurs in three stages [20, 21].

First, a specimen stretches uniformly on a macroscopic level so that no neck is formed. At the second stage the spherulitic structure transforms into a microfibrillar one. This transformation is usually accompanied by necking. The final stage involves the plastic deformation of the newly formed microfibrillar structure, which proceeds up to specimen breakdown [21].

We shall consider below the processes of defect formation and annihilation that occur during these three stages.

Even at small draw ratios lamellar crystals inside spherulites are deformed considerably. Deformation mechanisms are different in crystals, which are differently oriented with respect to the axis of extension. In the lamellae with chain axes inclined to the extension axis, chain slip and twinning can operate [21—23]. These processes are induced by shear stresses acting in the crystallographic planes parallel to chains. We can distinguish two basic modes of intralamellar slip. These are the slip along the chain axes and that in the transverse direction. Contributions of these modes to the deformation of a crystal depend on its orientation in a stress field which determines resolved shear stresses for longitudinal and transverse slip. The chain tilting, which is another possible deformation mechanism, proceeds via multiple slip along the chains so that is has similar nature.

The slip inside lamellar crystals should be realized by means of the dislocation motion [15, 16]. The driving forces acting on a dislocation are caused by shear components of an applied stress field. The plastic deformation of crystallites probably leads to an increases of the dislocation density inside them. This increase facilitates the breaking of lamellar crystals at larger elongations.

The twinning of polymer crystallites, by analogy with the low-molecular crystals, also should involve the motion of specific dislocations [24]. The twinning is usually considered to be responsible, for instance, for the formation of $a$-texture during drawing [20].

Interlamellar slip also contributes to the deformation of a spherulitic structure [20]. This slip is induced by shear stresses acting in the planes parallel to the large faces of lamellar crystals. In turn, the normal stresses perpendicular to these faces cause inelastic deformations of disordered layers being due to variations of the density of kinks in these layers.

The second stage of the deformation process usually starts at elongations of about 50%. This stage is distinguished by a drastic change of the polymer morphology. The transformation of original structure into the microfibrillar one is accompanied by the unfolding of chains. Accordingly, external 180° wedge disclinations, which model the elastic bending of the chains, disappear. In contrast, we suppose that internal disclinations, which correspond to gauche-conformers in the folds, persist through the transformation process. Therefore, the difference between the numbers of gauche-conformers of opposite sign in a single disordered interlayer of a microfibril should be prescribed by conformations of chain folds in original lamellar crystals.

The magnitude of this difference, $\Delta n_G$, calculated per unit chain in such a layer can be estimated on the basis of the following reasons. The energy associated with elastic bending of a folded chain in the crystallite becomes minimal when the fold contains a certain number of gauche-conformers. This number must ensure minimal deviation of the angle of conformational bend from 180° in a free chain. In the case of polyethylene and similar polymers each trans-gauche transition rotates the chain axis by 58° [5]. Hence, it is energetically most favorable to have three gauche-conformers per chain fold. Computer modeling of the chain folds in polyethylene supports the validity of our estimate: the number of gauche bonds varies between 1 and 4 for folds with different orientations in the lattice [25]. Thus, the mean number of excess positive or negative conformational defects per chain in an interlayer can be approximated as $\Delta n_G = 3$.

The data of IR-spectroscopy show that the total number of conformational defects of all types in a single interlayer is about $n_G = 10$ defects per chain [20], so that it is much larger than the number of excess defects. This difference is explained by the fact that kinks as topologically unstable defects can nucleate due to local internal rotations of backbone bonds caused by thermal fluctuations. Hence, the total concentration of conformational defects in the disordered phase after the formation of microfibrils is determined by physical properties of polymer molecules, as well as by the temperature and force conditions of drawing.

Disordered microfibrillar regions should contain not only conformational defects, but also external twist disclination loops. These crystallographic disclinations correspond to elastically twisted chain segments situated between neighboring crystallites, which hold them in a deformed state. Reneker and Mazur [10] showed that the existence of external twist loops in disordered interlayers is due to the twisting of chains, which usually accompanies the chain folding. This twisting is particularly evident in (110) fold planes typical of polyethylene crystals [10]. The planar zig-zag backbone must twist by approximately 90° in order to have correct orientation with respect to the lattice, which is necessary to reenter the crystal. Thus, the packing of macromolecules in the crystal lattice causes formation of external 90° twist loops in the folds. The amount of twist may change, but some persists through the transformation of initial structure into the fibrils. So, it is expected that a regular arrangement of chain twist boundaries is formed in a microfibril due to the localization of twist loops in transverse planes. These boundaries were first described by Reneker and Mazur [10].

In the next stage of drawing the deformation of the fibrillar structure takes place. The data of IR-spectroscopy indicate that in disordered layers the density of conformational defects decreases with increasing draw ratio [26, 27]. This phenomenon can be explained by the annihilation of defects of opposite sign within each layer. The most simple version of annihilation process consists in the elimination of a kink by means of internal rotation in the chain. The applied tensile stress induces a "driving force" for this process since a positive work is done by this stress during the elongation of a chain that accompanies kink elimination.

The reduction of the density of conformational defects due to their mutual annihilation inside disordered interlayers stops when only excess defects, which have the same sign within each chain segment between neighboring crystallites, are left in these interlayers. According to the above estimate these excess defects have appreciable density of about three defects per chain segment. They cannot be removed locally without breaking the skeletal bonds if the chain end is situated far from the disordered interlayer. In principle, it is possible to eliminate residual defects by their diffusion through a crystallite into neighboring disordered layers of a fibril, where they can annihilate with gauche-conformers of opposite sign. However, the stress field in a stretched polymer does not create any driving force for these conformational defects. In addition, the elastic interaction between defects located in the neighboring layers, according to the theory of disclinations [6, 28], appears to be negligible. Therefore, we expect that most excess defects in disordered intrafibrillar regions cannot be removed under usual conditions of drawing.

External twist loops existing in disordered layers of microfibrils also do not experience the action of mechanical driving forces during the drawing. However, a driving force of electric origin can act on such disclinations. This force appears in the case when polymer molecules carry a non-zero electric dipole moment per unit length, as, for instance, in ferroelectric polyvinylidene fluoride, and the drawing is carried out in a superimposed electric field. The motion of external twist loops along the chains is responsible for the transition of ferroelectric polymers into a macroscopically polar state [29, 30].

The main role in the deformation of polymers with fibrillar structure plays the mutual slippage of microfibrils [20]. For energetic reasons such a slippage cannot occur simultaneously over a large surface comparable with the lateral surface of a microfibril. Consequently, the slippage should involve nucleation and motion of the dislocation-like defects that model the border of incomplete interfibrillar shear. In the model case when a single microfibril is gliding relative to neighboring microfibrils, the carrier of plastic shear has the form of a prismatic quasi-dislocation loop enveloping certain cross-section of this microfibril (Fig. 5). The loop creates in this cross-section additional internal tensile stresses. Therefore, the probability of the rupture of a microfibril is increased in the vicinity of quasi-dislocation. The local tensile strength of a fibril should be lower in disordered interlayers since the chain segments between crystallites have different lengths. At moderate elongations the internal stresses can relax in these layers due to gauche-trans transitions that straighten the chains. In contrast, at large elongations exceeding a certain critical value the stress relaxation is suppressed because of the mechanically induced vitrification of disordered phase [31]. Accordingly, the formation of quasi-dislocations during the slippage of microfibrils can promote their rupture.

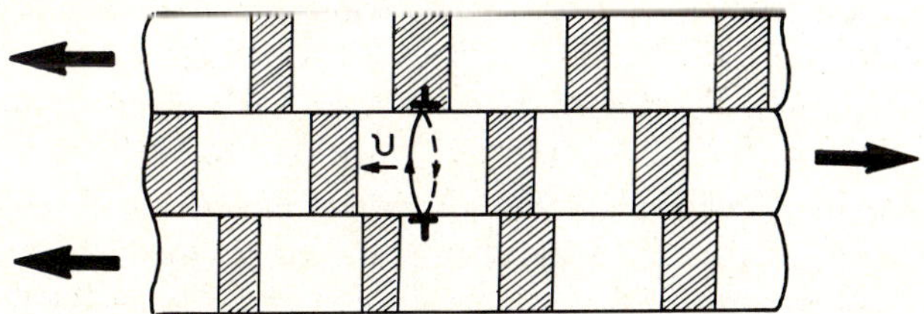


Fig. 5. The motion of a prismatic quasi-dislocation loop as a mechanism of mutual slippage of microfibrils during the drawing

The last stage of drawing is characterized by the appearance of numerous periodically spaced kink bands [32, 33]. These deformation bands are macroscopic defects of rotation type which usually have a thickness of the order of 1 μm.

It is well known that the kinking serves as the basic mechanism of plastic deformation in oriented polymers exposed to bending or compression along the orientation axis. In these situations the applied stresses produce a positive work during the kink band formation. In contrast, the applied tensile stress hinders the kinking in the case of drawing. Therefore, the formation of kink bands during the drawing should be regarded as a relaxation process. Marikhin et al. [33] suggested that this process is induced by internal compressive stresses of entropy origin which appear in a polymer owing to the breaking of stretched microfibrils.

The kink bands formed during the drawing have specific geometry (Fig. 6). They are situated in

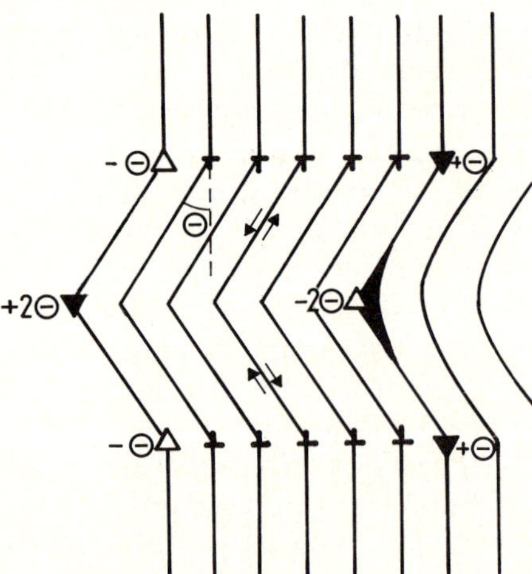


Fig. 6. Schematic illustration of a kink band formed in an oriented polymer at the last stage of drawing. Triangles denote the straight wedge disclinations located at the band fronts. The symbols ⊥ designate straight edge dislocations distributed on the band boundaries

planes perpendicularto the orientation axis and contain two differently reoriented zones [32, 33]. In contrast, ordinary kink bands formed under compression are inclined at an angle of about 45° to the orientation axis and include only one reoriented zone. These distinctions clearly indicate that during the drawing the kink bands nucleate under the action of compressive stresses, but not of shear stresses as in ordinary situation. On the other hand, the microscopic mechanism of kinking is the same in both cases; it involves the mutual slippage of microfibrils inside the band and their bending at the band boundaries [9].

The kink bands themselves are sources of internal stresses. The maximal stresses arise near the fronts of the incomplete kink bands that are blocked inside a sample [34]. These stresses can be calculated in continuum approximation if we take advantage of the models of kink bands worked out with the aid of the theory of dislocations and disclinations [34]. The analysis shows that the kink bands having the most simple shape are modeled by an ensemble of defects as described in Fig. 6. Here, triangles denote the straight wedge disclinations located at the band fronts. Uniform distributions of edge dislocations model the kink boundaries perpendicular to microfibrils. It should be noted that transverse tensile stresses are created in the vicinity of negative disclinations. Therefore, an interfibrillar crack must nucleate at the center of the right-hand front in the case when the bending angle $\Theta$ exceeds a certain critical value $\Theta^*$ depending on the transverse strength of oriented polymer. This leads to the exfoliation of the reoriented zone from adjacent material near the front of incomplete kink bands. Such an exfoliation is typical for the kink bands observed by Marikhin et al. in highly oriented polyethylene [32, 33].

The formation of kink bands during the drawing is very undesirable since the bending of fibrils at kink boundaries promotes their rupture [33]. However, in the presence of molecular kinks in disordered interlayers a certain part of the fibril bend will have conformational nature [35]. This conformational bend, which does not create tensile stresses inside the fibril, results from the disappearance of molecular kinks in the stretched part of interlayer and their nucleation in the compressed one (see Fig. 7). It should be noted that in a continuum approximation this rearrangement is equivalent to the formation of an effective wedge disclination loop surrounding the microfibril. The limiting angle $\Theta_m$ of a conformational bend calculated per single interlayer can be written as [35]:

$$\Theta_m = 2 \arctan\left(\frac{cpq}{2\delta}\right),$$

where $\delta$ is the transverse size of a microfibril, $c$ is the lattice period along the chains, and $p$ is the mean number of skeletal bonds in a chain segment between the neighboring crystallites. This equation shows that $\Theta_m$ strongly depends on the current density $q$ of molecular kinks in disordered interlayers. Accordingly, the limiting angle gradually decreases in the course of drawing owing to the reduction of the kink density $q$. The extent of stress relaxation that can be attained during the bend of microfibrils also decreases. Thus, the rupture of

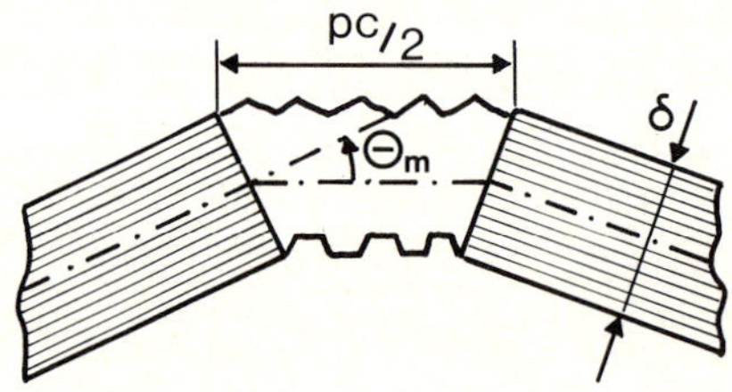


Fig. 7. Schematic illustration of a microfibril with the conformational bend localized within a single disordered interlayer

fibrils at the boundaries of kink bands becomes more probable at high draw ratios.

## 4. Conclusions

Application of the general theory of defects in solids to the studies of structural imperfections inherent in polymers leads to some new ideas about the behavior of these materials during the drawing.

From the topological stability of an elementary conformational defect we deduce that a number of gauche-conformers in disordered microfibrillar regions should be regarded as "excess" defects. These residual defects cannot be removed locally in contrast to the majority of conformational defects. The number of excess defects per chain is predetermined by the conformations of folds in lamellae of original unoriented polymer.

A mechanism of the mutual slippage of microfibrils during the drawing is proposed, which consists in the formation of quasi-dislocations and their subsequent motion along the fibrils. It is noted that quasi-dislocations as sources of internal stresses can promote the rupture of fibrils at high draw ratios.

The kink bands appearing in an oriented polymer during the last stage of drawing are described in continuum approximation. The disclination-dislocation model worked out permits to calculate the internal stresses induced by kink bands. On this basis it is possible to analyze the processes of microcracking caused by the formation of kink bands.

*Acknowledgement*

The author thanks Prof. V. A. Marikhin and Dr. L. P. Myasnikova for reading the manuscript and for valuable comments.

## References

1. Pechhold W, Blasenbrey S, Woerner S (1963) Colloid Polym Sci 189:14
2. Mc Mahon PE, Mc Cullough PL, Schlegel AA (1967) J Appl Phys 38:4123
3. Wunderlich B (1973) in Crystal Structure, Morphology and Defects, vol 1, Academic Press, New York,
4. Li JCM, Gilman JJ (1970) J Appl Phys 41:4248
5. Vladimirov VI, Pertsev NA (1984) In: Vladimirov VI (ed) Phys Techn Ins, Experimental Investigation and Theoretical Description of Disclinations, Leningrad (in Russian), p 37
6. Pertsev NA, Vladimirov VI, Zembilgotov AG (1989) Polymer 30:265
7. Pechhold W, Blasenbrey S (1970) Kolloid ZZ Polymere 241:955
8. Boyd RH (1975) J Polym Sci, Polym Phys Edn 13:2345
9. Pertsev NA, Romanov AE, Vladimirov VI (1981) J Materials Sci 16:2084
10. Reneker DH, Mazur J (1983) Polymer 24:1387
11. Reneker DH, Mazur J (1988) Polymer 29:3
12. Reneker DH (1962) J Polym Sci 59:S39
13. Reneker DH, Franconi BM, Mazur J (1983) J Appl Phys 48:4032
14. Harris WF (1974) Surf Defect Prop Solids 3:57
15. Frank FC, Keller A, O'Connor A (1958) Philos Magaz 3:65
16. Keith HD, Passaglia E (1964) J Res Natl Bur Std 68A:513
17. Holland VF (1964) J Appl Phys 35:3235
18. Lindenmeyer PH (1966) J Polym Sci (C) N15:109
19. Predecki P, Statton WO (1966) J Appl Phys 37:4053
20. Marikhin VA, Myasnikova LP (1977) Supermolecular Structure of Polymers, Khimiya, Leningrad (in Russian)
21. Peterlin A (1975) In: Ward IM (ed) Applied Science Publishers, Structure and Properties of Oriented Polymers, London
22. Allan P, Bevis M (1977) Philos Magaz A35:405
23. Allan P, Bevis M (1980) Philos Magaz A41:555
24. Dreimanis AP (1974) Mech Polym (Zinatne) N5:771
25. Petraccone V, Allegra G, Corradini P (1972) J Polym Sci (C) 38:419
26. Gafurov VG, Novak II (1970) Mech Polym (Zinatne) N1:170
27. Pakhomov PM, Shermatov M, Korsukov VE et al (1976) Vysokomol Soed 18A:132
28. Pertsev NA, Vladimirov VI (1983) Czech J Phys 33B:28
29. Reneker DH, Mazur J (1985) Polymer 26:821
30. Pertsev NA, Zembilgotov AG (1991) Sov Phys Solid State 33:165
31. Zhizhenkov VV, Egorov EA, Petrukhina TM (1973) Mech Polym (Zinatne) N3:387
32. Marikhin VA, Myasnikova LP, Pelzbauer Z (1981) Vysokomol Soed 23A:2108
33. Marikhin VA, Myasnikova LP, Pelzbauer Z (1983) J Macromol Sci Phys 22:111
34. Pertsev NA, Marikhin VA, Myasnikova LP, Pelzbauer Z (1985) Vysocomol Soed 27A:1438
35. Vladimirov VI, Zembilgotov AG, Pertsev NA (1989) Sov Phys Solid State 31:233

Author's address:

Dr. N. A. Pertsev
A. F. Ioffe Physico-Technical Institute
Russian Academy of Sciences
194021 St. Petersburg, Russia

**Progress in Colloid & Polymer Science** Progr Colloid Polym Sci 92:60—80 (1993)

# Orientation in networklike polymer systems. The role of extremum principles

H. G. Kilian, W. Knechtel, B. Heise, and M. Zrinyi*)

Abteilung Experimentelle Physik, Universität Ulm, FRG
*) Dept of Physical Chemistry, Technical University of Budapest, Hungary

*Abstract:* Large deformations of network-like systems are discussed. Such systems are characterized by the existence of equivalent subsystems of deformation isotropically linked with each other, operating as smallest structural units the shape of which is changed according to the law of an affine transformation. For describing large deformations of such "colloid systems" it is significant that the extremum principles of thermodynamics can be extended so as to allow a discussion of constrained equilibrium states. This has the consequence that strain invariant system elements in network-like polymer systems like real networks, glasses, filled rubbers or semicrystalline polymer system-elements are oriented so as to maximize the orientation entropy. This allows a concise thermodynamic interpretation. It is for example clear that in any constrained equilibrium state the maximum entropy is necessarily accompanied by an energy minimum. If internal constrained equilibrium is realizable even within very short periods of time these extremum principles should strictly be fulfilled. On the global level network-like systems should display universal features while the molecular structure comes into play on local scales. How both are interrelated and how they depend on the conditions of constrained equilibrium is discussed on the basis of representative examples.

*Key words:* Polyethylene — polymer glasses — deformation — orientation — extremum principles — irreversible thermodynamics

## Introduction

Polymer systems are in principle "crosslinked", for example, by chemical bonds, by crystals, by entanglements or interacting molecular units [1—10]. On deforming such "network-like" systems a defined transformation of a set of junctions or its equivalencies is enforced that depends uniquely on strain. The transformation is accompanied with defined orientation phenomena.

We assume that during deformation each network-like system passes constrained equilibrium states. It is then possible to apply thermodynamics of irreversible processes. A key step is here to define equivalent subsystems of deformation the shapes of which are transformed in an affine manner [11, 12]. Densely packed and linked together these subsystems ensemble constitute a quasi-continuum. Its deformation behavior is unique. Their orientation pattern does not depend on the absolute density or on the size distribution on the configuration of subsystems (scaling invariance).

Processes within subsystems determine the individual features of the stress-strain behavior. These processes depend on the molecular structure. In a permanent network, for example, chains of different lengths are individually stretched according to equipartitioning strain energy. Hence, long chains are extended much more than short ones. The mechanisms within the subsystems of network-like systems are assumed to satisfy the conditions of constrained equilibrium so that they should depend uniquely on strain.

The situation is illustrated in Fig. 1. A sketch of a "Kuhn-phantom network" (non-flucuating junctions) is drawn out in Fig. 1A [13, 14, 15]. Each

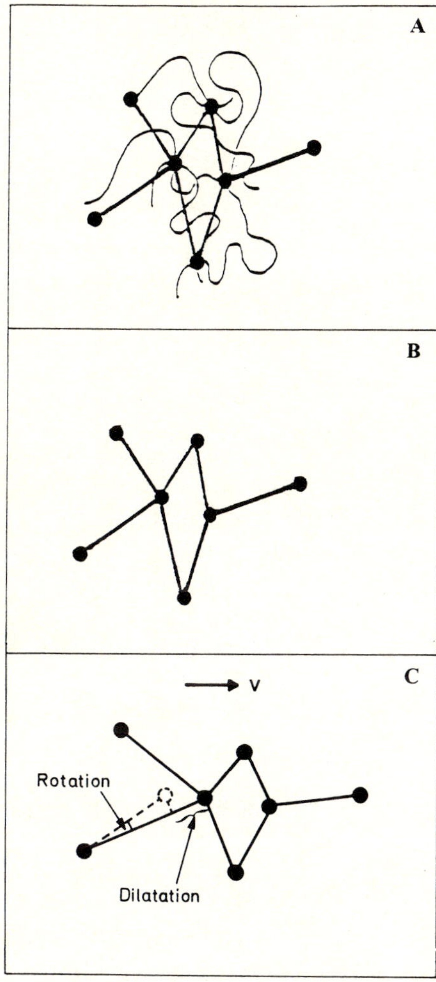


Fig. 1. A) Phantom network and its junction-to-junction connection network; B) the junction-to-junction connection network, C) affine deformation

junction-to-junction connection defines the symmetry axis of a subsystem of deformation. The orientation distribution of the junction-to-junction connections is given by the Kuhn-Kratky function independent on the density of crosslinks [16]. Non-affine conformational changes of chains guarantee an affine transformation of the junctions ensemble.

The global level of any network-like system is thus represented by the scheme in Fig. 1B. It includes "non-affine" processes in the subsystems. Network-like systems should althogether behave analogously even when the density of junctions is changed during deformation. It should in any case be possible to identify mechanisms of deformation by studying orientation phenomena on different levels.

To this end a general treatment is developed by describing constrained equilibrium states. With the aid of thermodynamics of irreversible processes generalized Gibbs functions are formulated. The thermodynamic extremum principles are then defined accordingly. This allows for example to give conditions every orientation distribution function should be submitted to. The above ideas are then proven and substantiated on the basis of representative examples.

## Thermodynamics

### *Gibbs-functions*

A thermodynamic system (homogeneous, single component system) obeys the fundamental equation [17—20]

$$dS(U,V,L) = \frac{1}{T}\,\{dU + pdV - fdL\}\ . \qquad (1)$$

$S$ is the entropy, $T$ the absolute temperature and $U$ the internal energy. $V$ defines the volume and $p$ the hydrostatic pressure. $f$ is the mechanic equation of state that defines the force field imposed onto the system from outside. $L$ is the extensive conjugate of the force. The last term in Eq. (1) gives the reversible strain energy under constant volume conditions. It was justified elsewhere [21] that even deformation-induced changes of volume are correctly described by using a Gibbs-function of the above type. The entropy is maximum at constant $U$, $V$ and $L$.

In most cases, we may now separate the entropy into two terms, the first related to orientation phenomena $S^{(\mathrm{network})}$ the second to all the other freedom $S^{(i)}$. We are then led to

$$S(U,V,L) = S^{(i)} + S^{(\mathrm{network})} = \max\,|_{U,V,L}\ . \qquad (2)$$

If $S^{(\mathrm{network})}$ itself is maximum any orientation distribution is well defined. Moreover, from thermodynamics it follows that the entropy-maximum is uniquely related to a minimum of strain energy

$$dU(S,V,L) = T(dS^{(i)} + dS^{(\mathrm{network})}) - pdV + fdL$$
$$U(S,V,L) = \min\,|_{S,V,L}\ . \qquad (3)$$

We lean from the integrability conditions

$$\left(\frac{\partial S^{(i)}}{\partial f}\right)_{S(\text{netowrk}),V,L} = \left(\frac{\partial L}{\partial T}\right)_{S(i),S(\text{network}),V}$$
$$\left(\frac{\partial S^{(\text{network})}}{\partial f}\right)_{S(i),V,L} = \left(\frac{\partial L}{\partial T}\right)_{S(i),S(\text{network}),V} \quad (4)$$

that entropy changes in a network-like system so that the kinetic energy is equipartitioned. This includes all strain-dependent entropic properties in a network-like system

$$\left(\frac{\partial S^{(i)}}{\partial f}\right)_{S(\text{network}),V,L} = \left(\frac{\partial S^{(\text{network})}}{\partial f}\right)_{S(i),V,L} . \quad (5)$$

An analogous relationship can be derived from the free-enthalpy

$$\left(\frac{\partial S^{(i)}}{\partial L}\right)_{T,p,S(\text{network})} = \left(\frac{\partial S^{(\text{network})}}{\partial L}\right)_{T,p,S^{(i)}} = -\left(\frac{\partial f}{\partial T}\right)_{p,L} \quad (6)$$

relating elastic and caloric properties.

Rewriting Eq. (1),

$$dS(U,V,L) = \frac{1}{T}\{dU^{(i)} + dU^{(\text{network})} + pdV - fdL\} \quad (7)$$

we are led to

$$\left(\frac{\partial U^{(i)}}{\partial f}\right)_{V,L,U(\text{network})} = \left(\frac{\partial U^{(\text{network})}}{\partial f}\right)_{V,L,U^{(i)}} = T\left(\frac{\partial L}{\partial T}\right)_{U(\text{network}),U(i),V} , \quad (8)$$

indicating that internal energy is equipartitioned as well.

*Maximum-entropy-orientation distribution*

We now apply a method that was introduced by Bower [27]. But we want to discuss the thermodynamic aspects behind it.

In an "orientational gas" the orientation entropy is maximum. Each segment occupies all accessible orientations with the same a priori probability. According to statistical thermodynamics [22—26] the entropy of the segments of a perfect fibre texture with an orientation distribution $\rho(\theta)$ may then be defined by the Boltzmann-relation

$$S_{\text{seg.}}^{(\text{network})} = -k_B \int \rho(\theta) \ln\rho(\theta) d\omega . \quad (9)$$

$k_B$ is Boltzmann's constant, $\theta$ is the angle between the axis of rod-like segment and the direction of extension. $d\omega$ is equal to

$$d\omega = 2\pi \sin(\theta) d\theta . \quad (10)$$

The orientation distribution $\rho(\theta)$ is normalized

$$2\pi \int_0^\pi \rho(\theta)\sin(\theta)d\theta = 1 . \quad (11)$$

In terms of Legendre polynomials $\rho(\theta)$ is represented by

$$\rho(\theta) = \sum_l a_l P_l(\cos(\theta))$$
$$a_l \begin{cases} = 0: & l = 2n+1 \\ \neq 0: & l = 2n \end{cases} . \quad (12)$$

In network-like systems $\rho(\theta)$ is uniquely defined by the degree of deformation. This is per definition clear if $\rho(\theta)$ describes the orientation distribution of the junction-to-junction connections. But it holds true for every other distribution, for example, for the orientation distribution of strain-invariant elements within the subsystems of deformation. In oriented networks orientation depends uniquely on all levels on the macroscopic strain. It is a significant matter that processes below the global level are not affine.

*Constrained equilibrium*

Even under large and rapid deformation sufficiently dense networks cannot be forced into large distances to equilibrium [28]. This emphasizes the use of thermodynamics of irreversible processes [29—32]. The above ideas can then be extended so as to describe constrained equilibrium states of matter. Deviation from absolute equilibrium may be looked at as originating by constraints. If these constraints are quasi-permanent the systems are supposed to occupy "constrained equilibrium" states. A network-like behavior is enforced, leading on the global level to an affine transformation, below this level to a non-affine transformation. Orientation entropy plays a substantial role.

Constrained equilibrium states can, in principle, be fully described. To this end, we define a complete set of additional hidden variables $\{\xi_k\}$ $(k = 1, \ldots, N)$ as constraint-parameters. The Gibbs function is then written accordingly

$$dS(U,V,L,\{\xi_k\}) = \frac{1}{T}\left(dU + pdV - fdL + \sum_{k=1}^{N} A_k d\xi_k\right)$$
$$A_k = T\left(\frac{\partial S}{\partial \xi_k}\right)_{U,L,V,\{\xi_j \neq \xi_k\}} . \quad (13)$$

$A_k$: affinity of the $k$th process.

The de Donder-affinity $A_k$ is conjugated to $\xi_k$. $A_k$ disappears in equilibrium, it characterizes the "distance from equilibrium" of the $k$th process. Because during relaxation each $d\xi_k$ is negative, the affinity $A_k$ must also be negative. This is necessary for guaranteeing relaxation-induced production of entropy.

So the entropy now shows a large number of maxima, each one related to the well defined and fixed set of hidden variables $\{\xi_k\}$

$$S = S(U, V, L, \{\xi_k\}) = \max |_{U,V,L,\{\xi_k = \text{const}_k\}} . \tag{14}$$

In the more specific form, written as

$$S^{(\text{network})}(U, V, L, \{\xi_k\}) = \max |_{U,V,L,\{\xi_k = \text{const}_k\}} . \tag{15}$$

According to the second law of thermodynamics we arrive at

$$S^{(\text{network})} = \max |_{U,V,L,\{\xi_k = \text{const}_k\}} < S^{(\text{network})}_{\text{equil}} . \tag{16}$$

In presence of constraints $\{\xi_k = \text{const}_k\}$ and constant $U$, $V$, and $L$, it is on the other hand clear that the number of accessible states is diminished. Yet, for a given set of constraints the orientation entropy is maximum while the internal energy $U$ is minimum

$$U = U(S, V, L, \{\xi_k\}) = \min |_{S,V,L,\{\xi_k = \text{const}_k\}} . \tag{17}$$

We come to the important result that in presence of quasi-permanent constraints the maximum entropy and the minimum internal energy are well defined. Entropy and internal energy are interrelated in the same manner as in classical thermodynamics.

*Orientation distribution under constraints*

From the above considerations it results that constraints modify the orientation distribution $\rho_\xi(\theta)$ [22—27]. Defining the moments $P_{i\xi}$

$$2\pi \int_0^\pi P_i(\cos(\theta))\rho_\xi(\theta)\sin(\theta)d\theta = P_{i\xi} \tag{18}$$

and maximizing the segmental entropy $S^{(\text{network})}_{\text{seg.}}$ we are led to

$$\rho_{\max\xi}(\theta) = \exp\left\{\sum_l \alpha_{l\xi} P_l(\cos(\theta))\right\} \tag{19}$$

with $\alpha_{l\xi}$ as Langrange multipliers. The actual values of the Langrange parameters depend on the density and the type of constraints.

It is now a crucial point that each orientation distribution can be approximated by using a limited number of moments. This number increases substantially with strain and with deviations from the monotonous Kuhn-Kratky-function (Table 1). To fully describe a pattern in highly oriented systems, one needs more then only the second and the fourth moments of the orientation distribution.

Table 1. Minimum number of required moments in dependence of the strain

| Kratky-Kuhn | |
|---|---|
| $\lambda$ | Required moments |
| 1.1 | $P_2$, $P_4$ |
| 1.5 | $P_2$, $P_4$, $P_6$, $P_8$ |
| 1.8 | $P_2$, $P_4$, $P_6$, $P_8$, $P_{10}$ |
| 2.0 | $P_2$, $P_4$, $P_6$, $P_8$, $P_{10}$, $P_{12}$, $P_{14}$ |

*Interesting consequences*

We take the incompressible elastic "quasi-continuum" as system of reference for characterizing in network-like systems. To this end, we define isotropic volume elements as subsystems. Each subsystem is linked to neighboring elements so that a network is build up (see Fig. 1). On deformation the whole ensemble is submitted to an affine transformation. Complicated non-affine processes run off within subsystems so that the junction-to-junction connection is rotated and extended accordingly. If we neglect the eigenvolume of subsystems the orientation distribution is given by the Kuhn-Kratky function

$$\rho(\theta) = \frac{\lambda^3}{1 + (\lambda^3 - 1)\sin^2(\theta))^{3/2}} \tag{21}$$

$$\int_0^{\pi/2} \rho(\theta)\sin(\theta)d\theta = 1 .$$

$\theta$ is the angle between symmetry axis (junction-to-junction connection) and draw-direction. The distribution is monotonous showing its maximum at $\theta = 0$ (Fig. 2). This Kuhn-Kratky function is the same for each subsystems network independent of its configuration and of the density of subsystems.

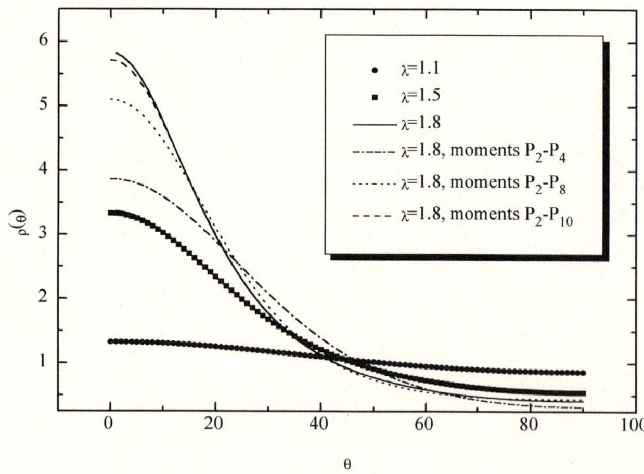


Fig. 2. The maximum-entropy orientation distribution (MED) compared with the Kuhn-Kratky-distribution. The number of moments used in the calculation is indicated with each curve

The second moment of the this distribution, $P_{2c}(\lambda)$, is equal to

$$P_{2c} = \frac{2\lambda^3 + 1}{2(\lambda^3 - 1)} - \frac{3\lambda^3}{2(\lambda^3 - 1)^{3/2}} \arctan \sqrt{\lambda^3 - 1} \,. \tag{21}$$

$P_{2c}(\lambda)$ has a concave curvature against the strain-axis [33]. Orientation increases essentially in the small-strain regime. This feature is typical and useful for a topological characterization.

It is now interesting that the Kuhn-Kratky function shows the maximum orientation entropy (Fig. 2). The orientation pattern is identical with the one of an ergodic system where each subsystem passes independently through all states of orientation.

The connected junction ensemble defines a quasi-continuum. It seems to satisfy on deformation maximum entropy and minimum energy conditions. One has to make sure whether this classification is justified.

## Network models

### *The Gaussean network*

Let us discuss the situation in the Gaussean network [7]. If the positions of junctions are not allowed to fluctuate they are moved on deformation according to the geometrical conditions of an affine transformation. Gaussean chains are equivalent: Strain energy and internal energy are equipartitioned among the network's chains. The internal energy

$$U = N\left(\frac{3}{2}\, k_B T\right) \tag{22}$$

does not depend on strain and is proportional to the number of chains $N$. The chains are energy-equivalent subsystems of deformation.

That the strain-energy is equipartitioned can be read from the equations

$$W = G\phi = nw$$

$$G = \frac{\rho R T}{M_c}\ ;\quad \phi = \frac{1}{2}\left(\sum_{k=1}^{3} \lambda_k^2 - 3\right) \tag{23}$$

$$n = \frac{N}{V}\ ;\quad w = k_B T \phi\ .$$

$G$ is the shear-modulus, $\rho$ the density and $T$ the absolute temperature. $\lambda_k$ is the strain parameter in the $k$th direction. $k_B$ is Boltzman's constant, $w$ the strain energy per chain. $n$ as the density of chains defines the density of subsystems of deformation.

### *The conformational and orientational gas*

The above model implicates [3, 34] that each junction-to-junction distance occupies independently all orientations. The crucial point is here that correlations due to the existence of junctions are neglected. The heuristic Kuhn-network is ergodic.

The phantom-network is an "ideal conformational and orientational gas" [10]. The Gaussean chains themselves correspond to a "phantom subsystem of infinite extensions without eigenvolume". This network satisfies the maximum entropy- and minimum energy principle. The orientation distribution of the junction-to-junction lines is identical with the universal Kuhn-Kratky function.

It is a crucial point that a Kuhn network with non-fluctuating positions of junctions seems to occupy just one of the many orientational configurations of the phantom network. The junction-to-junction lines in an incompressible elastic quasi-continuum then display the same orientation distribution.

Is there any physics behind this? Is an incompressible highly deformable elastic continuum a physically realistic idea? It is in any case clear that processes below the global level of the equivalent subsystems of deformation are, per definition, non-affine molecular processes. They cannot be described in a continuum approach. The situation is comparable with the "internal field problem" in a dielectric. The Lorentz sphere is the analogon of the subsystem of deformation. So, we need locally to discuss real processes like conformational changes of chains, like shear-sliding in crystals, like platzwechsel and slip. To achieve large deformations these processes must be activated. In the activated state the system should in fact be able to maximize the orientation entropy and to minimize the internal energy. This implicates sophisticated interrelations between these real processes and their continuum-like environment.

It is a question as to how we succeed in giving a reasonable description. For this and many other reasons we have to proceed to study real networklike systems.

Table 2. $P_2$ and $P_4$ from Langevin for different $N$ [35]

| $N = 50$ | | | $N = 100$ | | |
|---|---|---|---|---|---|
| $\lambda$ | $P_2$ | $P_4$ | $\lambda$ | $P_2$ | $P_4$ |
| 1.5 | 0.0066 | −0.0003 | 1.5 | 0.0033 | −0.0002 |
| 2 | 0.0145 | −0.0002 | 2 | 0.0072 | −0.0002 |
| 2.5 | 0.0243 | 0.0002 | 2.5 | 0.012 | −0.0001 |
| 3 | 0.0364 | 0.0009 | 3 | 0.0178 | 0 |
| 3.5 | 0.051 | 0.0021 | 3.5 | 0.0247 | 0.0002 |
| 4 | 0.0684 | 0.0038 | 4 | 0.0328 | 0.0007 |

*The Langevin network*

A first step in this direction is the Langevin network model [4]. It considers phantom-chains of finite length. The orientation distribution of the junction-to-junction lines is given by the Kuhn-Kratky function that controls the strain induced collective entropy loss on the global level.

From the plot in Fig. 3A it is to be seen that the mean orientation of chain segments in a network is always lower than the orientation of the junction-to-junction lines. The difference increases with increasing length of the mesh size. It is to be seen that the second moment of chain segments depends on strain in a clearly different manner as the junction-to-junction lines themselves [4].

It is interesting to recognize that it seems to be sufficient to know the second and the fourth moment [35] (Table 2) for approaching the real

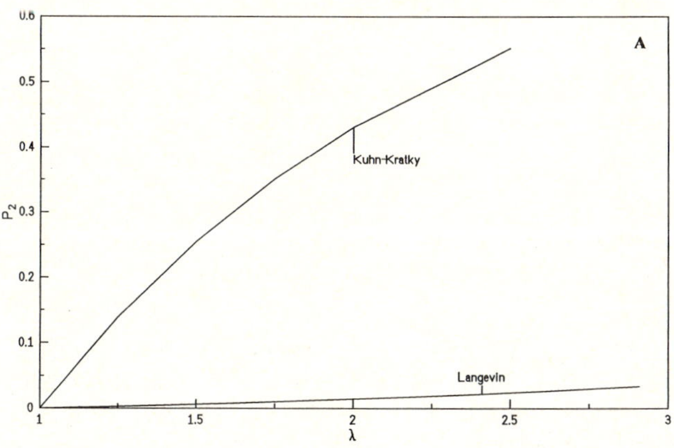


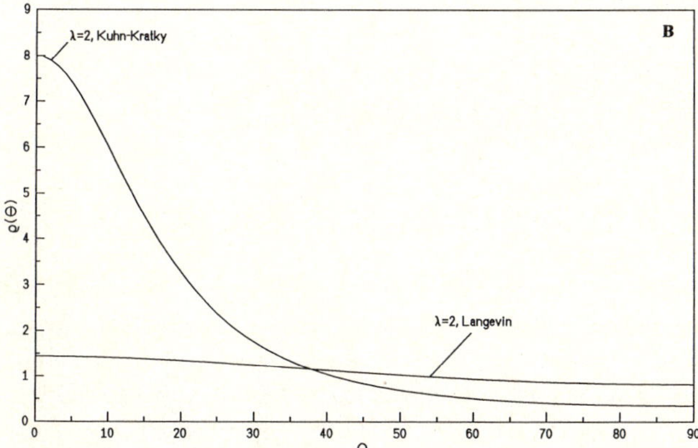


Fig. 3. A) $P_2(\lambda)$ for the different models. B) Maximum entropy orientation function for the Langevin- and the Kuhn-Kratky-model at $\lambda = 2$

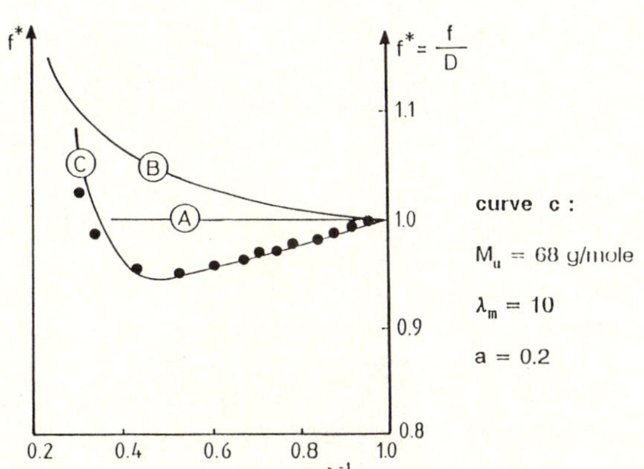


Fig. 4. Mooney-plot $f^*$ or NR at 363 K against $\lambda^{-1}$: a) Gaussean, b) Langevin, c) van der Waals

distribution as entropy maximum distribution (Fig. 3B). It is not easy to measure the segmental orientation distribution of networks in use, essentially in the small strain regime. This explains why the distribution is not directly deduced from experiments.

From the Mooney-plot in Fig. 4, one recognizes the well known observation that the Langevin model does not correctly give the stress-strain behavior in the whole range of strain [7, 36].

*The van der Waals network*

A full description of the thermo-elastic properties of networks is possible with the aid of a van der Waals modification of the Kuhn network [10]. The van der Waals equation of state reads

$$f = \frac{\rho RT}{M_c} D \left( \frac{1}{1 - \sqrt{\phi/\phi_{\max}}} - a\sqrt{\phi} \right)$$

$$\phi_{\max} = \frac{1}{2} \left( \sum_{k=1}^{3} \lambda^2_{\max,k} - 3 \right) ; \tag{24}$$

$$D = \frac{\partial \phi}{\partial \lambda}, \quad I_1 = \sum_{k=1}^{3} \lambda^2_{\max,k} .$$

In simple extension the maximum strain $\lambda_{\max}$ as the "first van der Waals parameter" is related to the chain-length parameter $y$ as the number of stretching-invariant units of the molecular weight $M_u$

$$y = \frac{M_c}{M_u} = \lambda^2_{\max} + \frac{2}{\lambda_{\max}} \approx \lambda^2_{\max} . \tag{25}$$

Strain-energy is equipartitioned among the chains [37]. For this reason the junctions have to fluctuate. The parameter $a$ as the "second van der Waals parameter" seems to account for the "softening" due to interactions coming about with these fluctuations [36].

The understanding of the stress-strain behavior for different types of strain is fairly good (Figs. 4, 5). We do not discuss the details.

*The stress-optical coefficient and the junctions fluctuation*

We demonstrate here that the anisotropy of different macroscopic properties must in general be related to different strain-invariant molecular segments. As a prominent example let us discuss the stress-optical law [7].

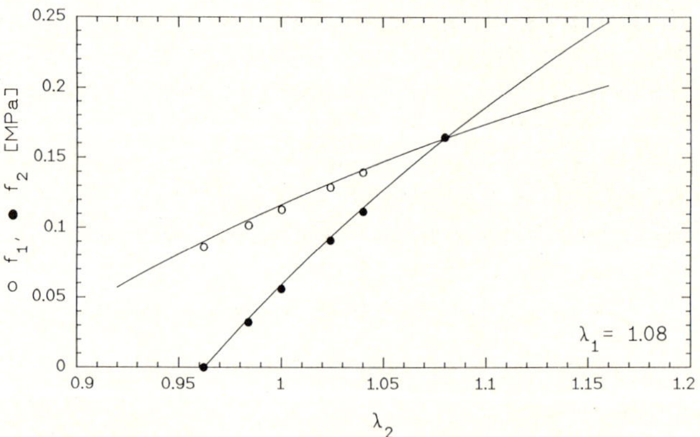


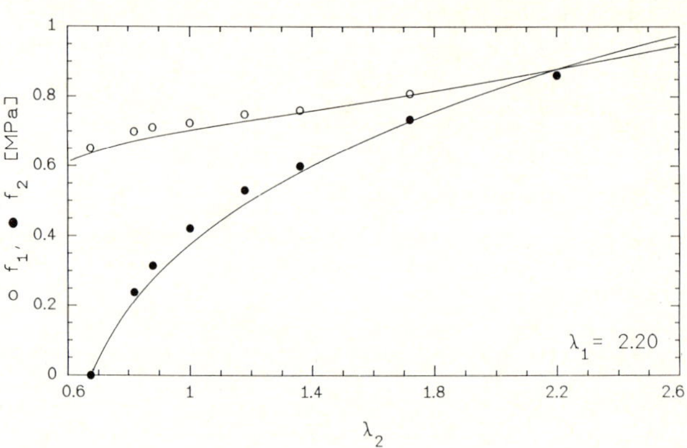


Fig. 5. Nominal stresses in equibiaxially deformed networks according to Kawabata et al. [72]; solid line computed ($I_1$ substituted by $I = \beta I_1 + (1 - \beta) I_2$; $\beta = 0.89$; $I_i$: strain invariants)

In terms of the van der Waals model any junctions fluctuations reduces the deformation entropy [36]. The effect should be proportional to

$$\text{correction term} \propto -aGD\sqrt{\phi} . \tag{26}$$

Because a stress-optical coefficient $C$ [7, 28] is always observed it is clear that junctions fluctuations should reduce orientation.

This is verified in the following manner. There are good reasons for assuming that the size of the stretching-invariant optical segment is not identical with the entropy-invariant segment. To describe this within the framework of the van der Waals approach, we define the molecular-weight of the "optically strain invariant segment" $M_{opt}$

$$\lambda^2_{max,opt} = \frac{M_c}{M_{opt}} \; . \tag{27}$$

The birefringence of an extended van der Waals-network may then be written as

$$\Delta n = \frac{\Delta n_0}{M_c}\left(\lambda^2 - \frac{1}{\lambda}\right) \cdot \left\{\frac{1}{1-\sqrt{\phi/\phi_{max,opt}}} - a_{opt}\sqrt{\phi}\right\} \tag{28}$$

$$\phi = \frac{1}{2}\left(\lambda^2 + \frac{2}{\lambda} - 3\right)\; ; \quad \phi_{max,opt} = \phi(\lambda_{max,opt}) \; .$$

Hence, the existence of a stress optical law demands that the interaction parameter $a_{opt}$ should be about the same as in Eq. (24) ($a_{opt} \cong a$)

$$C = \frac{\Delta n}{f\lambda} = \frac{\Delta n_0}{\rho RT}\,\frac{1/(1-\sqrt{\phi/\phi_{max,opt}}) - a\sqrt{\phi}}{1/(1-\sqrt{\phi/\phi_{max}}) - a\sqrt{\phi}} \approx \text{const.} \tag{29}$$

Treloar's experiments on natural rubber can in fact be described by using the parameters given in the capture of Fig. 6. Because the stress-optical coefficient of natural rubber decreases at larger strains the optically unit should be slightly smaller than the entropy-invariant segment ($M_{opt} < M_u$). Highly crosslinked polyethylene C.

Whatever the properties in networks are, they althogether depend uniquely on strain. Internal equilibrium is established.

### *Ergodicity*

In the classical sense networks are non-ergodic. But one has to be aware that subsystems are defined so as locally to fulfill the condition of ergodicity. The whole network runs through all possible configurations. They are characterized in the physical space by the existence of equivalent subcells. It is interesting to ask whether one might not define a "constrained ergodicity" reduced due to the presence of permanent junctions. Clearly, this is a confinement which should influence fluctuations in a unique dependence on the junctions configuration.

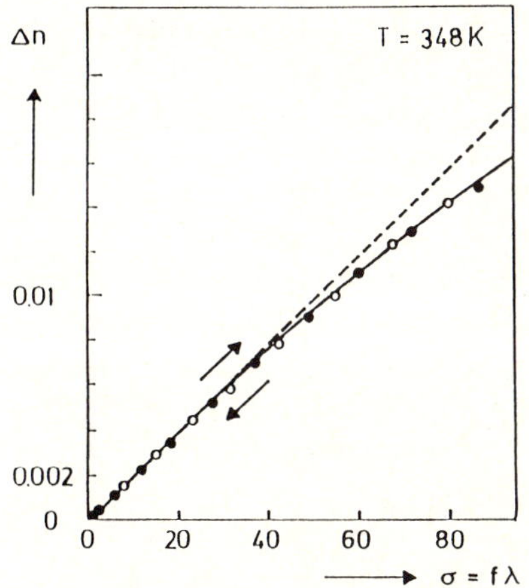


Fig. 6. Stress against birefringence for NR according to Treloar [7]. The solid line is computed ($\lambda_{max}$ = 9.9, $a$ = .28, $M_u$ = 68 g mol$^{-1}$; $M_{opt}$ = 66 g mol$^{-1}$; $\Delta n_0$ = 0.003)

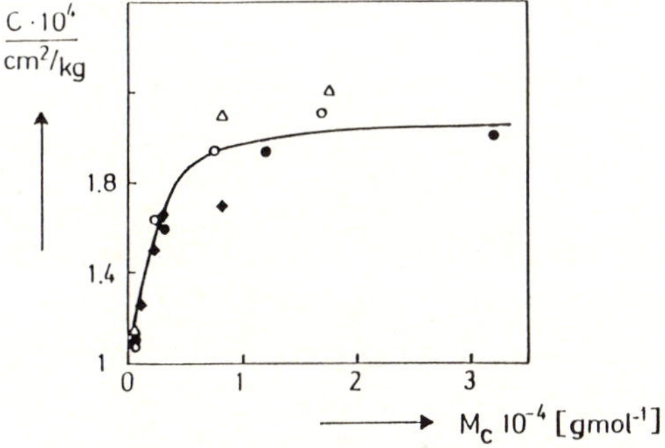


Fig. 7. Stress-optical coefficient of polyethylene against the mean molecular weight (solid line computed; $a$ = 0.3, $M_u$ = 14 g mol$^{-1}$, $M_{opt}$ = 56 g mol$^{-1}$ according to [7]

## Examples and discussion

### *The glassy continuum*

It is now a question of outstanding interest whether polymer glasses can be considered as network-like systems. First of all it is observed that polymer glasses deform irreversibly up to large strains. Orientation seems to be controlled by a set of strain-invariant hidden variables [39]. The strain energy of this network-like system may thus be characterized by Eq. (23) at least at large strains. The size the of shear-modulus $G$ characterizes the density of that constraints that are frozen so as to constitute a quasi-permanent "physical network"

$$G = \frac{G_0}{M_{\text{subsystem}}} \ . \tag{30}$$

In a glass this density of quasi permanent constraints is expected to be extremely high. The size of subsystems of deformation should be represented by smallest aggregates of molecular segments which could undergo an affine deformation. The density of the subsystems should be inversely proportional to their total mass $M_{\text{subsystem}}$.

There are good reasons for assuming that $M_{\text{subsystem}}$ is in the order of magnitude of a monomer unit [42]. From a discussion of the stress-strain behavior of glassy polycarbonate it follows that the junction-to-junction vector should have a weight of $M_{\text{subsystem}} = 85 \text{ g mol}^{-1}$. They are anisotropic units that move like rigid rods embedded in a solid environment with a liquid like short range order. Strain- or stress activated non-affine slip- or platzwechsel processes have to run off so as to allow an affine transformation of the shape of the subsystems. This in in good accord with neutron-scattering experiments as performed by Dettenmaier and Kausch [43]. Heats measured during quasi-isothermal extension can also be interpreted as orientational entropy effects due to very short-chain segments [39].

*The orientation distribution*

The glass is a frozen defect-saturated system. Segments linked in a very short "glassy network" become strongly oriented, rearranging themselves without producing measurable defects. Because of the smallness of the strain-invariant segments their eigenvolume and shape come into play. Both effects may be accounted for by using the Oka-type function [40, 41]

$$\rho(\theta) = \frac{\lambda^a}{(1 + (\lambda^a - 1)\sin^2(\theta))^{3/2}} \tag{31}$$

$$a = 3\,\frac{k^2 - 1}{k^2 + 1} \ ; \ k = \frac{a}{b} \quad \begin{matrix} a\text{: width} \\ b\text{: length} \end{matrix} \ .$$

$a$ is defined by the shape parameter $k$. Only, aggregates with an anisometric shape suffer orientation (Table 3). The Kuhn-Kratky function corresponds to the limit $k \to \infty$.

Table 3. $P_2$ from Oka for different shape-factors $k$ at $\lambda = 2$

| $k$ | $P_2$ |
|---|---|
| 2 | 0.2636 |
| 4 | 0.3853 |
| 6 | 0.4116 |
| 8 | 0.4211 |
| 10 | 0.4225 |
| Kuhn-Kratky | 0.431 |

If the glass-typical constraints are constant $\{\xi_k = \text{const}_k\}$ smallest segments behave like Kuhn-Kratky-Oka-rods. The birefringence $\Delta n$ is consequently defined by

$$\Delta n = \Delta n_0 P_{2c}^a$$

$$P_{2c}^a = \frac{2\lambda^a + 1}{2(\lambda^a - 1)} - \frac{3\lambda^a}{2(\lambda^a - 1)^{3/2}} \arctan\sqrt{\lambda^a - 1}$$

$$\Delta n_0 \begin{cases} \text{stretching invariant} \\ \text{optical segment} \end{cases} \ . \tag{32}$$

It is now possible to correctly compute the second moments in coldly drawn polycarbonate (Fig. 8) by assigning the form anisometry parameter to $k = 1.92$. The finding of [44] that $P_4$ is a unique function of $P_2$ is in agreement with our description (Fig. 10). The entropy maximum principle allows now to deduce the whole orientation distribution (Fig. 9).

These results prove that segmental orientation in glasses depends uniquely on strain. According to Eq. (16) this is guaranteed only if the constraints $\{\xi_k = \text{const}_k\}$ are strain-invariant. This was just the assumption we used in describing the energy balance in coldly drawn polycarbonate [39].

The maximum entropy orientation distribution of segments in a glass can apparently be achieved by platzwechsel activated during deformation at constant constraints $\{\xi_k = \text{const}_k\}$. With stopping deformation one of the large numbers of possible constrained equilibrium configurations is frozen in. Because platzwechsel are activated for a short period of time only ergodicity even in its constrained version as discussed above cannot be proven at all. Yet numerous configurations of a glassy network should be equivalent. This guarantees that the

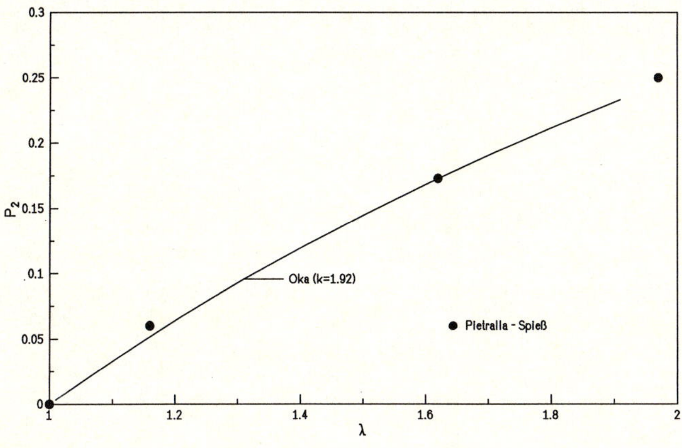


Fig. 8. Polycarbonate: $P_2(\lambda)$ of polycarbonate according to [44], solid line theoretical

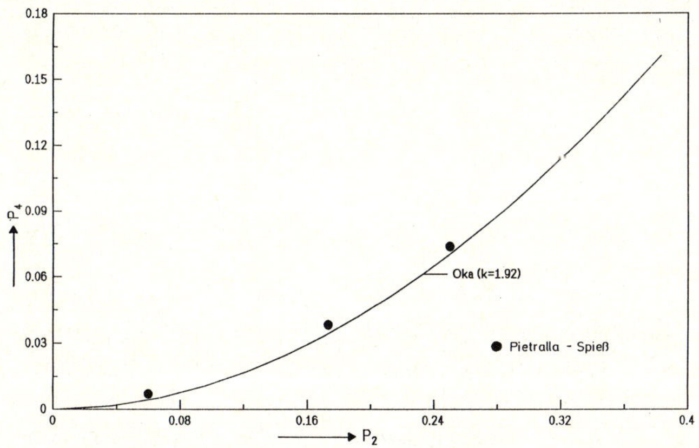


Fig. 9. $P_2(\lambda)$ against $P_4(\lambda)$ of polycarbonate according to [44], solid line theoretical

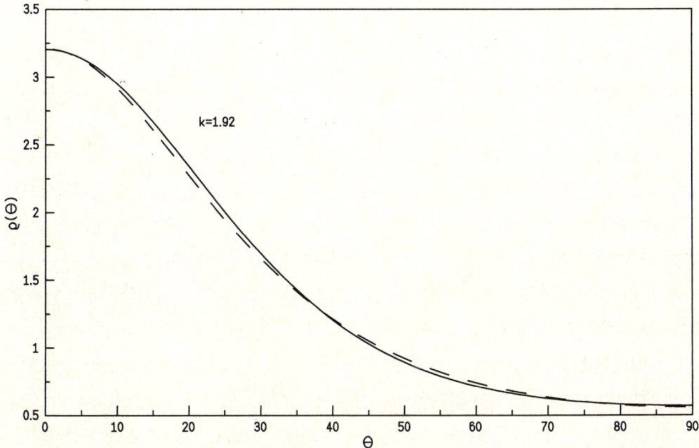


Fig. 10. $\rho(\theta)$ of polycarbonate computed at $\lambda$ 3 2, its maximum entropy representation (solid line) and the Oka-function (dashed line) with $k = 1.92$

freezing process is, under defined conditions, macroscopically reproducible. The above hypothesis may in addition be taken as indirectly verified by the finding that the segmental orientation is distributed so as to maximize the entropy.

A glass may indeed be considered as defect saturated network-like system. Strain energy should then be minimized. This hypothesis is not easy to prove.

*Rigid rods in a rubbery matrix*

It would be fine if we could prove our conception directly. To this end, we inserted rodshaped colloid particles into a rubber matrix [45]. After having a strained sample fixed with a special preparation method the orientation of rods or clusters of rods can be observed in electron microscope pictures (Fig. 11).

At larger extensions one observes an orientation pattern as depicted in Fig. 12. Rods or clusters or rods with a different form anisometry are oriented

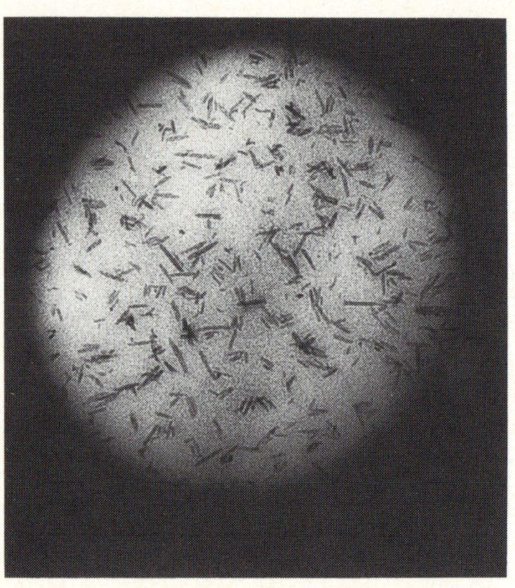

a)

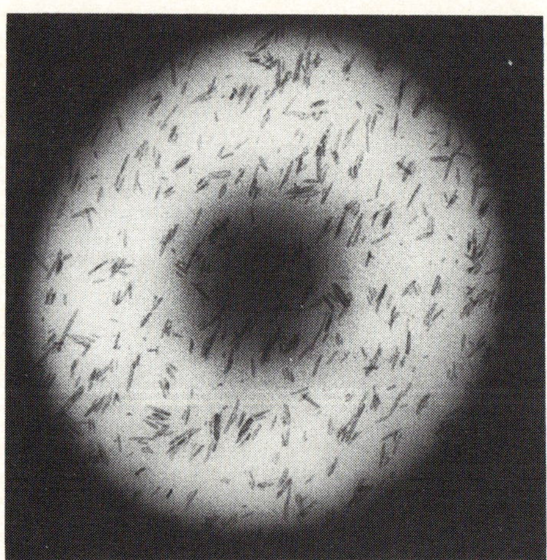

b)

Fig. 11. Electron-micrographs of iron-oxide complex rodlike primary particles according to [70]; A) unstrained, B) $\lambda = 4.1$

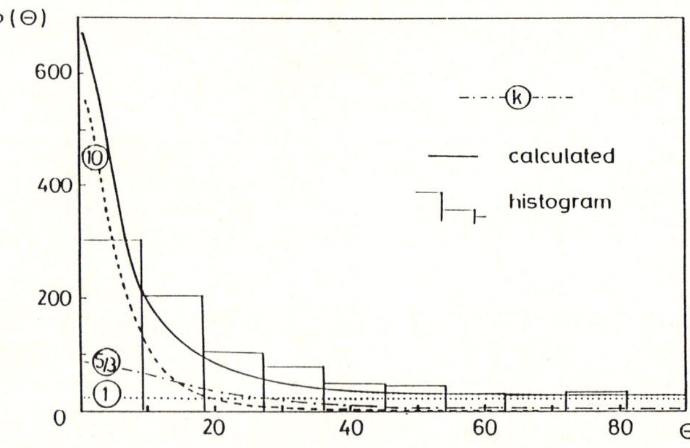


Fig. 12. Histogram ($\lambda = 4.1$) deduced from the electron micrographs; $\rho(\theta, k_i)$ of orientation of rods or clusters with anisometry parameter $k$ indicated with each curve computed with Oka-components as indicated. Solid line: the total orientation distribution

differently, each type according to its anisometric shape. Clusters with an isotropic shape cannot become oriented at all.

The orientation distribution of each anisometric filler aggregate can be described with the aid of the Oka-function (Eq. (32)) by assigning $k$ to the values deduced from the electron-microscope pictures. Hence, rods or aggregates of different anisometry behave like independent components, each maximizing its orientational entropy.

It is not known how rapid these rods move according to their micro-Brownean motions. It is, on the other hand, important that a large number of network-chains are in contact with each colloid-particle (more than 1000). Collective forces excerted onto each filler particle or filler-particle aggregate should therefore bring it into its minimum energy position. If each rigid rod operates like a segment in a glass, each should be oriented in the direction of the junction-to-junction line of the subsystems involved.

In the present case subsystems of deformation are apparently constituted by the filler particles or by clusters together with their next environment. There are therefore subsystems of different size and different shape. The rubbery regions where the clusters are embedded into suffer a complicated "non-homogeneous" deformation. To establish the conditions of affinity for each subsystem "slip-processes" at the filler particle's surface should happen to occur. It is significant that the "plasto-elastic" cooperation is optimized. The whole ensemble displays then an orientational pattern that maximizes the orientational entropy. That the strain energy is minimized is proven by finding rods or anisotropic aggregates always in the direction of the junction-to-junction lines ("ideal plastic deformation mode").

It is not expected that the rods or clusters move so rapidly during deformation that each one independently occupies each orientation. The system displays, nevertheless, a maximum entropy orientation distribution thus showing the characteristics of an ergodic ensemble. This is again only understandable if filler particles or clusters are activated to move during extension. This has to happen so as to bring the whole system into one of the many constrained equilibrium configurations. Hence, large heterogeneous deformation in filler loaded rubberlike systems is highly cooperative.

*Semicrystalline polymers*

In semicrystalline polymers a "crystal network" is formed [1, 11, 12, 33, 46—48]. Neighboring crystals are linked by tie-molecules, many of them in parallel arrangements (Fig. 13). The existence of a network is clearly proven by finding fibre symmetry in largely extended samples [46, 49]. Strain-energy in the rubbery regions is not equipartitioned among the chains. These network-like systems show reinforcement [49]. Processes like crystallographic sliding, chain slip and melting and recrystallization come into play. These processes reduce reinforcement.

It is an interesting question how to define the equivalent subsystems of deformation.

*The subsystems of deformation*

Crystals undergo crystallographic deformation processes like twinning [50—55] or shear-sliding [33]. If the extremum principles hold true, conservative sliding planes should only come into action and minimize the deformation energy. Because of the finite number of these crystallographic planes very specific constraints are developed. They should give the c-axis-orientation distribution typical features.

If crystals operate as solid particles the intrinsic strain in the non-crystallized regions $\lambda_i$ should be larger than the macroscopic strain $\lambda$. The system is reinforced whereby reinforcement is continuously reduced with strain because the crystal ensemble undergoes plastic deformation.

Above the glass-transition the non-crystallized regions are rubbery regions. Yet, non-crystallized layers between crystal lamellae cannot behave like rubber essentially because of not being able to contract so much in lateral directions. Hence, shear sliding, twinning, [50—55] melting and recrystallisation, chain-slip of very short chains together with shearing of fibrils [33, 47, 48, 56—60] must be enforced. It is thus clear that only a larger assembly of clusters can affinely be transformed. The subsystems ensenble should then be transformed under extremum conditions so as to achieve the maximum orientational entropy- and the minimum energy.

An unstrained subsystem is likely to show a spherical shape. For low-density polyethylene one observes in the electron-microscopic picture

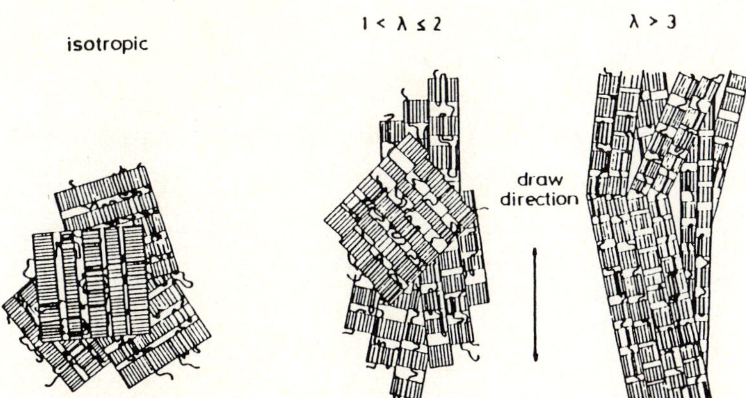


Fig. 13. Cluster-structure transformation enforced by uniaxial extension

($R \approx 4$ µm) circular cross-sections of small spherulites. In strained samples ellipsoids appear (Fig. 14). This is in good accord with observations by [61]. The mean ratio of the principle axis satisfies the condition of affinity. But the large axis is always nearly in direction of the stress. Hence, it appears that every spherulite is an aggregation of subsystems finally transformed into fibrils [57].

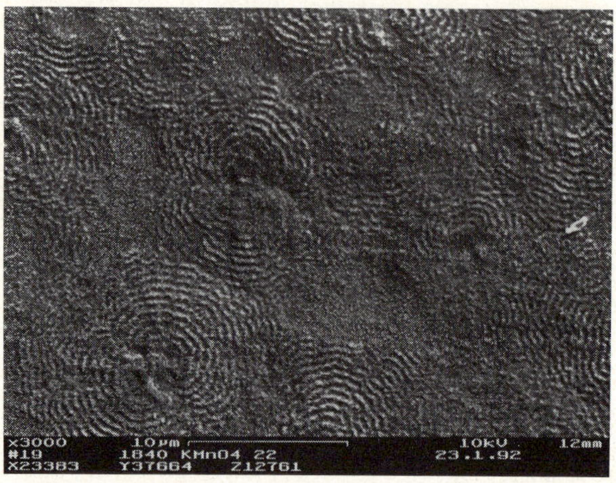


A)

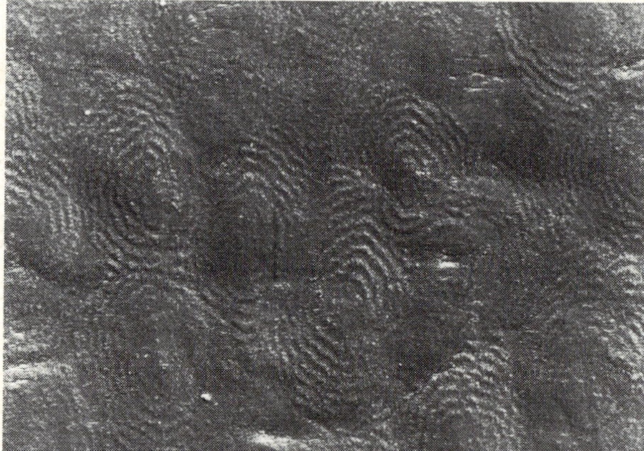


B)

Fig. 14. Electron-scanning micrographs of stained LDPE A) unstrained, B) $\lambda = 2$

*The mode of mechanical cooperation*

Because crystals and non-crystallized layers are basically linked in series the force exerted onto both should on average be identical.

$$\langle f^{\text{matrix}} \rangle = \langle f^{\text{crystals}} \rangle = f^{\text{macroscopic}} = f \; . \tag{33}$$

If the equation of state of the rubbery regions is known it is possible to compute stress-strain pattern. In this paper we use the van der Waals-equation of state.

*Primary parameters of the colloid structure*

The properties of a crystal network depend, of course, on the degree of crystallinity and the mean thickness of crystals. Because a large number of crystals and matrix elements cooperate during deformation within the subsystems of deformation it is sufficient to know the mean thickness of the crystals only. To study low density polyethylene (LDPE) as eutectoid copolymer is therefore rather promising because the primary parameters are thermodynamically well defined.

Extended-$CH_2$-sequence mixed crystals (ESMCs) are formed. Defective boundaries are developed. The best understanding of mechanical properties is achieved if rather the whole ESMC is assumed to operate as a rigid and active filler. The boundaries behave like "rigid amorphous regions" in the sense of Wunderlich [62].

Crystallization in LDPE is uniquely determined by the molar fraction of the non crystallizable units (nc-units) $x_{nc}$

$$x_{\text{nc}} = \frac{n_{\text{nc}}}{n_c + n_{\text{nc}}} \; , \; x_c + x_{\text{nc}} = 1 \; . \tag{34}$$

We rest content with using a random distribution of branchings so that we have the fraction of $CH_2$-sequences given by

$$x_y = x_{\text{nc}} x_c^{y(T)-1} \; . \tag{35}$$

The molar fraction of crystallized $CH_2$-sequences defines then the "crystallinity $w_p$"

$$w_p = \sum_{y(T)}^{\infty} y(T) x_y = x_c^{y(T)-1} (y(T) x_{\text{nc}} + x_c) \; . \tag{36}$$

$y(T)$ characterizes the thickness of the smallest ESMCs which are just stable at the temperature $T$. $w_p$ is substantially larger than the x-ray crystallinity $w_c$ where the core of the extended-sequence mixed lamellae matters only ($T = 295$ K, $w_p = 0.76 > w_c = 0.42$ [63]). The mean thickness of the ESMCs is then written as

$$\langle y_p \rangle = \frac{x_c}{x_{\text{nc}}} + y(T) \; . \tag{37}$$

The primary structure parameters are altogether uniquely defined by the chain structure. How they depend on temperature is also known. Selective- and consecutive melting of ESMCs of different thickness is described with the aid of thermodynamics of eutectoid copolymers [64—66].

*The crystal-network*

A "crystal-network" operates similar like a filled rubber. Crystals are rigid and active multifunctional fillers.

It is reasonable to relate the effective chain length in the network to the mean distance $D_{cryst}$ between crystals. Because of finding on average each crystal surrounded by non-crystallized regions, we have

$$D_{cryst} = \langle y_p \rangle \frac{1 - w_p^{1/3}}{w_p^{1/3}} . \tag{38}$$

The mean effective chain length may then simply be written as

$$\lambda_{max}^2 = y_{chain} = a \langle y_p \rangle \frac{1 - w_p^{1/3}}{w_p^{1/3}} . \tag{39}$$

The parameter $a > 1$ expresses that the contour length of the effective chains is larger than the mean distance between crystals. $a$ can be invariant only if chain-folds and entanglements are distributed so as to always satisfy the relationship as defined in the above equation. With the assumption that chains in the unstrained system are fully relaxed the maximum strain $\lambda_{max}$ is defined by

$$\lambda_{max}^2 + \frac{2}{\lambda_{max}} \cong \lambda_{max} = \sqrt{y_{chain}} . \tag{40}$$

Slippage and pulling out of chains as well as melting and recrystallization should increase the mean effective mean chain length. In accordance with Eq. (40), we formulate [63]

$$\lambda_{max,\lambda} = \sqrt{a \langle y_p \rangle \frac{1 - (w_{p/\lambda})^{1/3}}{(w_{p/\lambda})^{1/3}}} . \tag{41}$$

The transformation of the superstructure in a unique dependence on strain is behind this mean-field description. Hence, Eq. (41) describes a strain-induced modification of constraints in semicrystalline low-density polyethylene. Looking at it in this way, we have to understand deformation in network-like systems showing a density of junctions which decreases in a well defined dependence onto the macroscopic strain [63].

Due to having the rubbery regions "overdrawn" every semicrystalline system is reinforced. The mean strain in rubbery regions $\lambda_i$ was then written as [11, 12]

$$\lambda_i = \frac{\lambda - u_p}{1 - u_p} \; ; \; u_p = \left( \frac{w_p}{\lambda} \right)^{1/3} . \tag{42}$$

The exponent 1/3 indicates that the subsystems are quasi-isotropically linked in their environments. In the low strain regime the rubbery layers should be "overdrawn" ($\lambda_i \gg \lambda$) as more as the crystallinity $w_p$ gets larger. In the "Bueche-limit" without any plastic deformation of crystals $u_p$ should be equal to $u_p = w_p^{1/3}$. Plastic deformation and intrinsic reshuffling change the mean geometrical shape of the crystal ensemble. It is for this reason that $\lambda_i - \lambda > 0$ decreases with increasing strain. To account for these effects two fractions have been introduced in Eq. (42)

$$\begin{aligned} &\frac{1}{\lambda} : \text{ rigid fraction} \\ &1 - \frac{1}{\lambda} : \text{ plastic fraction.} \end{aligned} \tag{43}$$

Due to always having the rigid fraction $1/\lambda$ of ESMCs the whole ensemble is not transformed according to the law of affinity. Compensation is achieved by the softer rubbery regions within each subsystem.

Because of having $\lambda_i$ uniquely related with $\lambda$, the plastic strain parameter of the crystal ensemble $\lambda_c$ is obtained from

$$\lambda = w_p^{1/3} \lambda_c + (1 - w_p^{1/3}) \lambda_i \tag{44}$$

to be equal to

$$\lambda_c = \frac{\lambda - (1 - w_p^{1/3}) \lambda_i}{w_p^{1/3}} . \tag{45}$$

*The stress-strain equation*

Despite the sophisticated structure of the crystal network we apply the van der Waals model. The interpretation is thus given in terms of a homogeneous network. We have then the mechanical equation of state [10, 63]

$$f = \frac{\rho RT}{M_u \lambda^2_{\max,\lambda}}$$

$$\cdot\ D(\lambda_i) \left\{ \frac{1}{1 - \sqrt{\phi(\lambda_i)/\phi(\lambda_{\max,\lambda})}} - a\sqrt{\phi(\lambda_i)} \right\}$$

$$\approx \frac{\rho RT}{M_u \lambda^2_{\max,\lambda}} \frac{D(\lambda_i)}{1 - \sqrt{\phi(\lambda_i)/\phi(\lambda_{\max,\lambda})}} \ . \tag{46}$$

Multi-functional active fillers like crystals should not allow large fluctuations. The relationship at the bottom of the above equation should therefore be a good approximation.

*Comparison with experiments*

The stress-strain pattern of LDPE in dependence of temperature is fairly well described with the aid of the Eqs. (41), (42) and (46) (Fig. 15). The parameters are collected in Tab. 4. They are identical with the ones in the unstrained system. Effects like strain-induced melting or crystallization are neglected [7, 67, 68].

Table 4. LDPE: parameters for calculation

| LDPE | | | | | | |
|---|---|---|---|---|---|---|
| $T$/°C | 20 | 30 | 40 | 50 | 60 | 70 |
| $y_{min}$ | 28 | 30 | 32 | 35 | 38 | 42 |
| $w_p$ | 0.76 | 0.745 | 0.72 | 0.67 | 0.64 | 0.59 |
| $y_p$ | 57 | 59 | 61 | 64 | 67 | 71 |

It is also possible to describe the stress-strain patterns of high density polyethylene (HDPE) at different temperatures. Here, we have folded chain crystals the formation of which depends on crystallization history so that the parameters of the colloid-structure parameters have to be deduced from experiments.

With the thermoelastic equations of state, derived for constrained equilibrium states, it is possible to compute the differential heats exchanged during a quasi-isothermal stretch (Fig. 16) [69]. With exception of the small strain regime the experiment is satisfyingly reproduced. The ratio of the differential heats of deformation related to the stress approaches rapidly the value of one [69]. This is typical for systems where work exchanged is identical with [19]

$$W_{\text{Net.}} = -T\Delta S_{\text{Net.}} \ . \tag{47}$$

LDPE behaves in the regime beyond $\lambda > 1.5$ like an entropy-elastic network, the structure of which is continuously changed. The effective chain-length increases substantially with strain (Fig. 17) indirectly characterizing the dramatic transformation of the colloid structure.

From the calculations it is clear that equilibrium in the rubbery regions is established. At each strain the entropy is maximum while the strain-energy is minimum. Whether now the crystal ensemble itself satisfies these extremum conditions may for example be checked by studying the orientation distribution of crystals.

*Energy losses*

Before we proceed to do this let us discuss the elementary situation in network-like systems which suffer a strain-induced softening. We interpret relation (46) as equation of state that gives the nominal force at a distinct value of $\lambda_{\max}(\lambda)$. Because the density of junctions decreases with strain the work indicated with the hatched area in Fig. 18 must be dissipated. The differential heats can be deduced from the Eq. (47) with the aid of

$$f_s = \mathrm{T}\left(\frac{\partial f}{\partial T}\right)_{p,L} = -\left(\frac{\partial Q}{\partial L}\right)_{p,T} \ . \tag{48}$$

We arrive at the representation as shown in Fig. 18. The strain-induced softening of the cluster network is due for example to pulling chains out of crystals, to shear sliding of crystal blocks or to melting and recrystallization of crystals. The elastic strain energy stored at $\lambda$ is then simply written as

$$W_\lambda = -G_\lambda\{(\ln(1-\eta) + \eta)\, 2\phi_{\max,\lambda}\} \ ;$$

$$(\lambda_{\max} = \text{const.}) \ . \tag{49}$$

The total energy put into the system is computed by the integration of the actual nominal force (Eq. (46))

$$W_{\text{tot}} = \int_{\lambda'=1}^{\lambda} f(\lambda')d\lambda' \ . \tag{50}$$

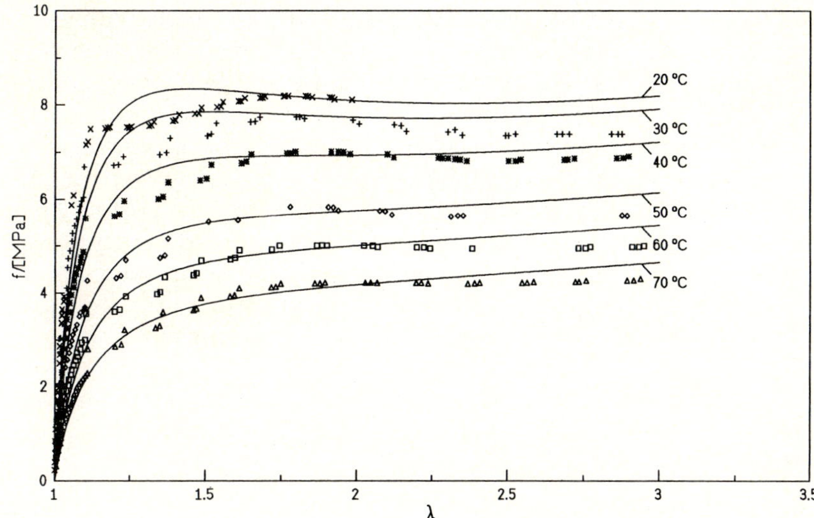


Fig. 15. Stress-strain curves of LDPE at the temperatures as indicated according [11] (parameters see Table 4)

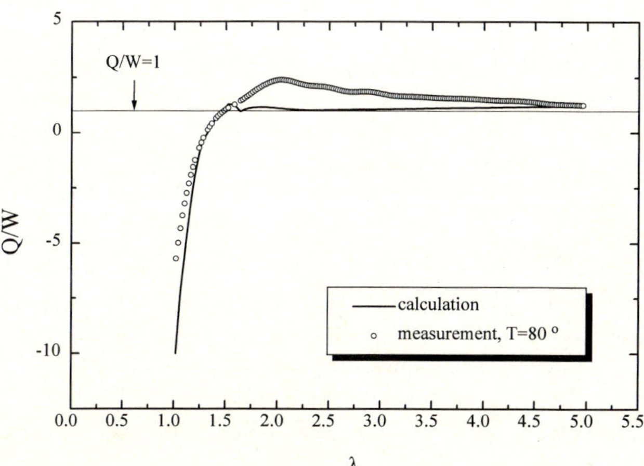


Fig. 16. The ratio of the differential heats $Q/W$ of LDPE against strain at the temperature as indicated according to [69]

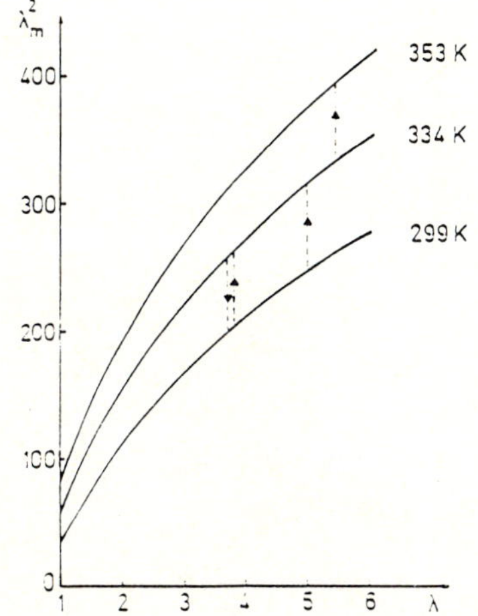


Fig. 17. $\lambda_m^2$ of LDPE against strain at indicated temperatures

The relative fraction of energy dissipated is then equal to

$$\eta_{\mathrm{diss},\lambda} = \frac{\Delta W_\lambda}{W_{\mathrm{tot}}} = \frac{W_{\mathrm{tot}} - W_\lambda}{W_{\mathrm{tot}}} \quad . \tag{51}$$

This quantity is plotted against the strain for two temperatures in Fig. 19.

We learn from this example that the strain-induced annihilation of constraints is well defined in strain. The "distance to equilibrium" is reduced. For a cluster network the heuristic equilibrium state is the isotropic melt. Because quasi-permanent junctions are released with strain entropy is produced by relaxation. We arrive at the very interesting balance equation

$$\begin{aligned} T(\Delta S_{\mathrm{net},\lambda} - \Delta S_{\mathrm{diss}}) &= 0 \\ T\Delta S_{\mathrm{net},\lambda} &= \Delta W_\lambda \ . \end{aligned} \tag{52}$$

By releasing constraints the chains themselves attain very rapidly a "more coiled conformation". The heat bath is the sample. It would under adiabatic

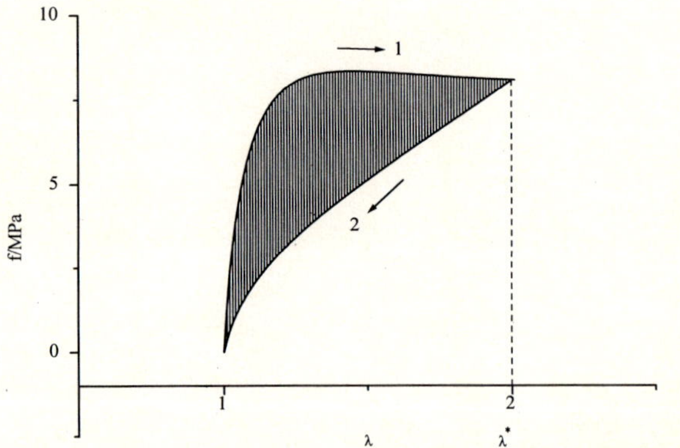


Fig. 18. The nomial forces of a cluster network with $\lambda_{max}(\lambda)$ stretched up to $\lambda^*$ (1) and the unloading curve of the permanent-network with $\lambda_{max}$ = const. (2)

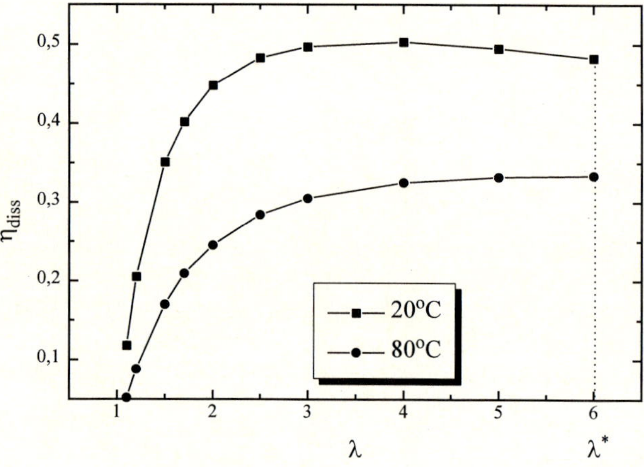


Fig. 19. Fraction of dissipated energy ($\eta_{diss}$) of LDPE at temperatures as indicated

conditions always keep the same temperature independently of which amount of the elastic work is just dissipated. Such an "isothermal relaxation" holds true in the exceptional case of a purely entropy elastic system. It is thus allowed to compute the heat transfer in a quasi-isothermal experiment with the aid of Eq. (48). We are led to conclude that crystal-cluster networks operate — with exception of the small strain regime — like purely entropy-elastic networks. Strain-induced network softening runs so as to produce entropy under given circumstances in unique dependence on strain. There are, of course, time effects, the study of which would inform about relaxation of the global structure.

The question is now whether constrained equilibrium is achieved at each strain.

*Orientation distribution*

At not too large strains the *c*-axis orientation distribution of polyethylene shows a more over less pronounced maximum in the range of 20—40 degrees [1, 2, 33]. Shear-sliding in the crystals enforces typical deviations from an affine transformation. But these deviations decrease at larger strains so that the maximum is increasing shifted into the drawing direction.

It is shown in Figs. 20 and 21 that the measured *c*-axis distributions are reproduced with the aid of the maximum entropy principles. One has only to know the three experimental moments $P_{2c}$, $P_{4c}$ and $P_{6c}$. This is in accord with Bower [27]. The *c*-axis-distributions of LDPE (Fig. 20) and HDPE

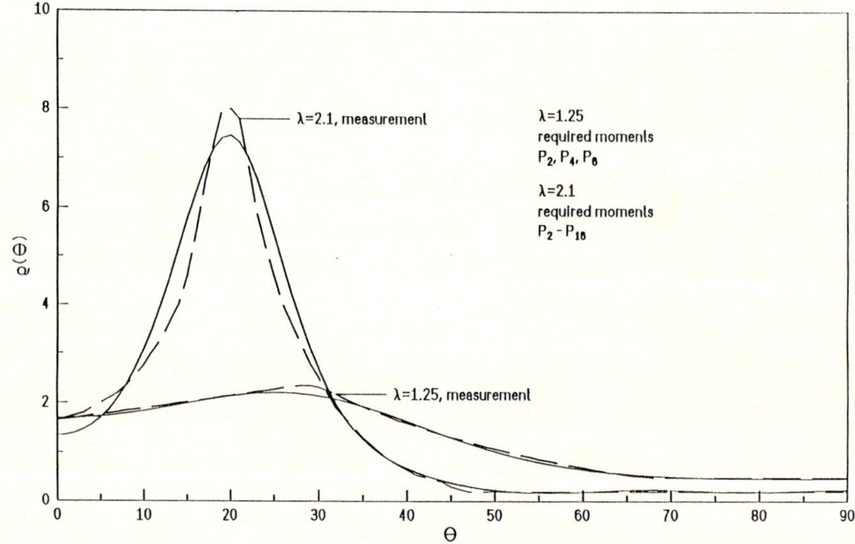


Fig. 20. $\rho(\theta)$ of the *c*-axis of LDPE at room-temperature and for strains as indicated [47]; dashed lines: entropy-maximum distribution, computed with the $P_i$s indicated

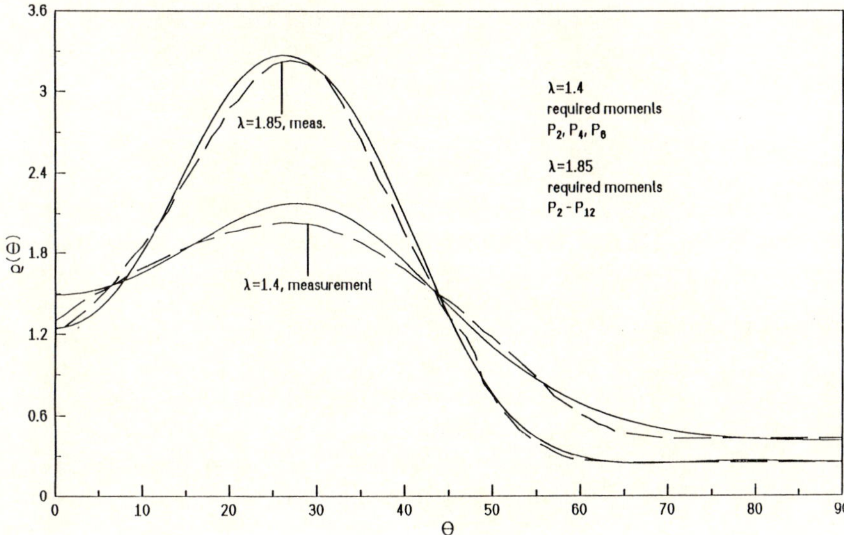


Fig. 21. $\rho(\theta)$ of the *c*-axis of HDPE at room-temperature and for strains as indicated [47]; dashed lines: entropy-maximum distribution: computed with the $P_i$s as indicated

(Fig. 21) are shown together with their maximum entropy distribution.

It is in evident that the *c*-axes do not display the same orientation as the symmetry axis of the subsystems of deformation given by the Kuhn-Kratky distribution. The deformation processes within the rubbery regions and the ESMC-ensemble must be different. But if subsystems of deformation do exist both must be coupled to each other so that it is guaranteed that the shape of the subsystems is affinely transformed.

### *A new transformation*

It is clear that plastic deformation of the ESMCs ensemble is submitted to crystallographic conditions like the one that slip processes should run off in lattice planes. Such constraints control the deformation essentially in the small-strain regime. Whatever occurs at largest strains the deformation should approach an affine behavior. Putting that together, we define the following transformation

$$\sin\theta = \lambda \sin\theta' \frac{\cos\theta^*}{\cos\theta} \lambda_2$$

$$\lambda_2 = \lambda^{\frac{1}{2}\left(\frac{a_s}{\lambda}\sin\theta' + 1 - \frac{1}{2}\right)} \tag{53}$$

$$\theta^* = \frac{1}{\lambda}\left(1 - \frac{2\theta'}{\pi}\right)\theta' + \left(1 - \frac{1}{\lambda}\left(1 - \frac{2\theta'}{\pi}\right)\right)\theta\,.$$

This leads to the orientation distribution function

$$\rho(\theta';\lambda) = \xi \frac{\lambda(\cos\theta^* + \tan^2\theta'\cos\theta^*) - \tan\theta'\sin\theta^*\left(1 - \frac{4\theta' + 2\theta}{\pi}\right) + \ln\lambda\,\frac{a_s}{2\lambda}\cos\theta}{\cos\theta + \lambda\tan\theta'\sin\theta\left(1 - \frac{1}{\lambda}\left(1 - \frac{2\theta'}{\pi}\right)\right)} \tag{54}$$

$$\xi = \frac{\lambda\tan\theta'\cos\theta^*\,\lambda_2^2}{\sin\theta'}$$

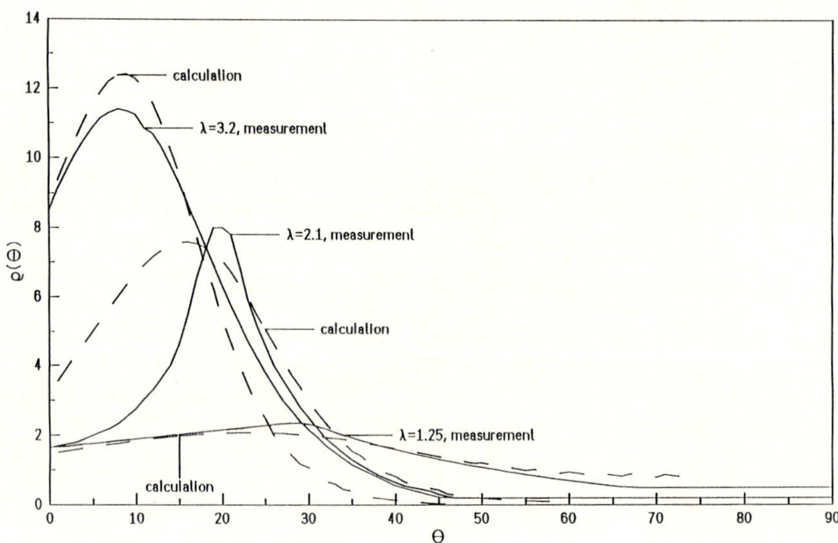


Fig. 22. $\rho(\theta)$ of LDPE at the strains as indicated with each curve; Solid lines theoretical

If $\lambda_c$ replaces $\lambda$ the orientation distribution of the $c$-axis of LDPE-ESMCs is fairly well described (Fig. 22). Systematic discrepancies are observed in the range of $\lambda \approx 2$. The parameters used in the calculations are collected in Tables 4 and 5.

From Eq. (44), one learns that sliding and rotation of crystals control processes in the small strain regime. $a_s$ is here larger than one. Crystal clusters having their $c$-axis perpendicular to the direction of stress become possibly rotated so as to bring the $c$-axis rapidly into the draw direction. Amorphous layers within these clusters should then strongly be sheared in lateral directions. Crystals with the $c$-axis parallel to the draw direction should, on the other hand, undergo shearing as defined in the shear-model [33]. The "crystallographic deformable fraction" is reduced proportional to $1/\lambda$. At elevated strains melting and recrystallization might increasing help to bring the $c$-axis into the subsystems symmetry-axis.

Table 5. Parameters for calculation of $\rho(\theta)$ of LDPE with [54]

| $\lambda$ | $\lambda_c$ | $a_s$ |
|---|---|---|
| 1.25 | 1.106 | 4 |
| 2.1 | 1.8 | 3 |
| 3.2 | 2.8 | 1 |

*The a- and b-axis orientation*

In orthorhombic crystals, we have the conservative sliding system constituted by the set of (110)-

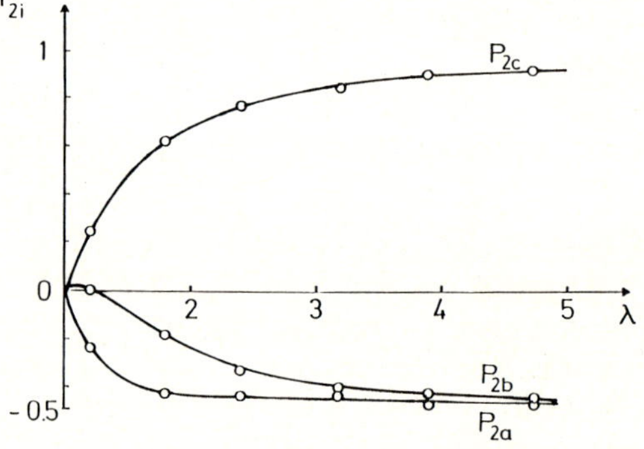


Fig. 23. $P_{2c}$, $P_{2a}$ and $P_{2b}$ of LDPE aginst $\lambda$ as deduced from WAXS-measurement [71]; the solid lines are computed [63]

planes. The atomistic density is here maximum. The $a$- and $b$-axis distribution is fairly well computed (Fig. 23) if one assumes that $c$-axis slip within the (110)-planes runs so that the normal of these planes, the $c$-axis and the direction of stress are in plane [63]. The same result is obtained for HDPE.

It is apparently correct that a crystal network deforms under minimum strain-energy and maximum entropy conditions. One might believe that crystals as rigid entities cannot move like elements of an ergodic system. Yet, one has to keep in mind that chain-segments within the ESMCs are laterally exchanged all the time [63—65]. Mixing in ESMCs is bound to such a "dynamics" within the relatively thin lamellae. These processes seem to be sufficient for minimizing the density of conservative sliding planes during deformation, minimizing therewith the strain-energy. This implies that the orientation entropy of the crystal ensemble is maximized at each strain. At the end the crystal ensemble occupies one of the possible configurations of the large set which the system would run through under constrained equilibrium conditions.

## Final statements

Thermodynamics can be extended so as to describe constrained equilibrium states. This is most adequate for describing deformation in network-like polymer systems. All processes, also the ones which enforce orientation, depend uniquely on the macroscopic strain.

The global level in very different network-like systems seems clearly to be defined by a densely packed and a priori isotropically linked assembly of subsystems of deformation. The orientation distribution of the symmetry axis of subsystems is universal and given by Kuhn-Kratky-Oka function. This distribution shows the maximum of orientational entropy of a homogeneously deformed continuum.

It is shown that strain-invariant chain segments in oriented polymer glasses realize an orientation distribution of the Kratky-Oka type. At constant constraints platzwechsel seems to be activated during extension within very short periods of time. One of the many different constrained equilibrium configurations is attained and frozen in. The liquid-like short-range order seems to allow sufficient "non-affine" molecular rearrangements within the subsystems themselves.

In a strained rubber form-anisometric larger rigid rods or clusters of such rods take their energy minimum orientation by being oriented in parallel direction of the symmetry axis of the subsystem. It is very likely that here a single rod or cluster defines together with the environment a subsystem of deformation. The orientation distribution corresponds to the maximum entropy distribution whereby clusters of different form anisometry are individually adjusted.

Intra-subsystem mechanisms must in any case guarantee the affine transformation of the shape of the subsystems even in crystal networks. Crystals and rubbery regions are transformed very differently but are strictly coordinated. The cluster-network is increasingly softer with increasing strain. Each of the components, ESMC-ensemble or rubbery regions, is transformed during deformation so as to achieve constrained equilibrium. The orientational entropy is maximized while the strain energy is minimum. Plastic deformation of crystals runs across conservative sliding planes only.

Network-like systems are able to dissipate strain energy at constant temperature by directly transferring entropy of orientation into the heat bath. Far above the glass temperature this dissipation is determined by the rate with which constraints are released. In a polymer glass not so much energy can be directly dissipated.

This paper shows that network-like systems seem altogether to be able to realize constrained equilibrium even if the dynamics is restricted to short periods of time, for example, during the activated states of coldly drawn glasses. Orientation reveals in any case that the entropy maximum principle are satisfied in the very different network-like systems. Yet it is for example difficult to explain straight forwardly how the *c*-axes of the crystals in a cluster network optimize their orientational entropy. Any explanation should consider local processes running within the subsystems. These local processes may mechanically be activated so as to allow for necessary reshuffling.

These local processes are cooperatively linked to the subsystems network. Here the transformation is universal as long as hidden variables are strain-invariant or as long as they are uniquely disposed with strain. The systems "optimize themselves" by forming an adequate set of densely packed quivalent subsystems of deformation. Local processes running off within the equivalent subsystems of deformation may be very specific, they may be strain activated, but they must in any case operate so as to guarantee that constrained equilibrium states are realized even under the most exotic conditions.

A deeper theoretical understanding of the many suggestions is desirable.

## References

1. Ward IM (1975) Structure and Properties of Oriented Polymers. Applied Science Publishers, London
2. Ward IM (1971) Mechanical Properties of Solid Polymers. Wiley Interscience, London
3. Flory PJ (1953) Principles of Polymer Chemsitry. Cornell University Press

4. Kuhn W, Grün G (1942) Kolloid Zeitschr 101:248
5. James H, Guth E (1943) J Chem Phys 11:455
6. James H, Guth E (1947) J Chem Phys 15:669
7. Treloar LRG (1975) The Physics of Rubber Elasticity. Clarendon Press, Oxford
8. Kramer O (1992) In: Mark JE, Erman B (eds) Elastomeric Polymer Networks. Prentice Hall, Englewood Cliffs
9. Graessley WW (1975) Macromolecules 8:186
10. Kilian H-G (1981) Polymer 22:209
11. Mayer J, Schrodi W, Heise B, Kilian H-G (1990) Acta Polymerica 41:363
12. Mayer J, Heise B, Kilian H-G (1988) 7th Intern Conf Deformation, Yield and Fracture. Cambridge 65:1
13. Kuhn W (1936) Kolloid Zeitschr 76:258
14. Kuhn W (1939) Kolloid Zeitschr 87:3
15. Wall FT (1942) J Chem Phys 10:485
16. Kratky O (1933) Kolloid Zeitschr 64:213
17. Haase R (1963) Thermodynamik der irreversiblen Prozesse. Steinkopff, Darmstadt
18. Callen HB (1960) Thermodynamics. John Wiley & Sons, New York
19. Ambacher H, Kilian H-G (1992) In: Mark JE, Erman B (eds) Elastoemric Polymer Networks. Prentice Hall, Englewood Cliffs
20. Godovsky YK (1986) Adv Pol Sci 76:31
21. Keller JU (1977) Thermodynamik der irreversiblen Prozesse, Pt 1. de Gruyter, Berlin
22. Haken H (1990) Synergetik. Springer, Berlin
23. Jaynes ET (1957) Phys Rev A 106:620
24. Levine RD, Tribus M (1979) The maximum entropy formalism. MIT Press, Cambridge
25. Berardi R, Spinozzi F, Zannoni C (1992) J Chem Soc Faraday Trans 88:1863
26. Berne BJ, Pechukas P, Harp GD (1968) J Chem Phys 49:3125
27. Bower DI (1981) J Pol Sci Pol Phys Ed 19:93
28. Kraus V, Kilian H-G, v Soden W (1992) Progr Coll & Pol Sci 90:27
29. Llorente MA, Andrady AL, Mark JE (1981) J Pol Sci Pol Phys Ed 19:621
30. Meixner J (1954) Zeitschr Naturforschg 9a:654
31. Kilian H-G, Vilgis T (1984) Coll & Pol Sci 262:691
32. Kilian H-G, Vilgis T (1984) Coll & Pol Sci 262:696
33. Heise B, Kilian H-G, Pietralle M (1977) Prog Coll Pol Sci 62:16
34. Mark JE, Erman B (1988) Rubberlike Elasticity-A Molecular Primer. Wiley & Sons, New York
35. Nobbs JH, Bower DI (1978) Polymer 19:1100
36. Zrinyi M, Kilian H-G, Horkay F (1989) Colloid Polym Sci 267:311
37. Winkler G, Reineker P, Schreiber M (1983) Europhys Lett 8:493
38. Rehage G, Schäfer EE, Schwarz J (1971) J Angew Makromol Chem 16:231
39. Koenen JA, Heise B, Kilian H-G (1989) J Pol Sci Pol Phys Ed 27:1235
40. Oka S (1939) Kolloid Zeitschr 86:242
41. Peetz L, Krüger JK (1986) Colloid Polym Sci 264: 1010
42. Schubach HR, Nagy E, Heise B (1981) Colloid Polym Sci 259:789
43. Dettenmaier M, Kausch HH, Nguyen TQ, Higgins JS (1984) Bull Am Phys Soc 29:415
44. Weeger R, Pietralla M, Peetz L, Krüger JK (1988) Colloid Polym Sci 266:692
45. Zrinyi M, Kilian H-G, Dierksen K, Horkay F (1991) Markomol Chem Macromol Symp 45:205
46. Adams WW, Young D, Thomas EL (1986) J Mater Sci 21:2239
47. Balta Calleja FJ, Kilian H-G (1985) Colloid Polym Sci 263:692
48. Kilian H-G, Balta Calleja FJ (1988) Colloid Polym Sci 266:29
49. Mayer J, Schrodi W, Heise B, Kilian H-G (1990) Acta Polymerica 41:363
50. Pietralla M (1974) Dissertation, Abt Experimentelle Physik. University Ulm
51. Groves GW, Kelly A (1970) Crystallography and Crystal Defects. Longman
52. Cahn RW (1954) Twinned Crystals, in Adv in Physics 3:12
53. Jawson MA, Dove DB (1960) Acta Cryst 13:232
54. Allan P, Crellin EB, Bevis M (1973) Phil Mag 27:127
55. Young RJ (1983) Introduction to Polymers. Chapman and Hall, London
56. Kinloch A, Young RJ (1988) Fracture Behaviour of Polymers. Elsevier, London
57. Peterlin A (1971) J Mat Sci 6:490
58. Peterlin A (1979) Ultrahigh Modulus Polymers. Cifferei A, Ward IM (eds) Appl Sci Publ, London
59. Godovsky YK, Bessonova NP, Mironova NN (1986) Colloid Polym Sci 264:224
60. Ward IM (1985) Adv Polym Sci 66:81
61. Attenburrow GE, Bassett DC (1979) Polymer 20:1313
62. Wunderlich B (1973) Macromolecular Physics. Academic Press, New York
63. Schrodi W (1991) Dissertation, Abt Experimentelle Physik. University Ulm
64. Kilian H-G (1986) Progr Colloid Polym Sci 27:60
65. Rosenberger B, Asbach GI, Kilian H-G, Wilke W (1988) Makromol Chem 189:2627
66. Glenz W, Kilian H-G, Klattenhoff D, Stracke F (1977) Polymer 18:685
67. Mandelkern L (1964) Crystallization of Polymers. Mc Graw Hill, New York
68. Conradt RNJ (1990) Diplomarbeit, Abt Experimentelle Physik. University Ulm'
69. Borst T (1991) Diplomarbeit, Abt Experimentelle Physik. University Ulm
70. Haas W (1991) Diplomarbeit, Abt Experimentelle Physik. University Ulm
71. Brenner M (1989) Diplomarbeit, Abt Experimentelle Physik. University Ulm
72. Kawabata S, Matsuda M, Tei K, Kawai H (1981) Macormolecules 14:154

Authors' address:

Prof. Dr. H. G. Kilian
Abt. Experimentelle Physik I
Universität Ulm
Albert-Einstein-Allee 11
89081 Ulm, FRG

**Progress in Colloid & Polymer Science** Progr Colloid Polym Sci 92:81—102 (1993)

# Chain extension and orientation: Fundamentals and relevance to processing and products

A. Keller and J. W. H. Kolnaar[1])

H. H. Wills Physics Laboratory, University of Bristol, Bristol, United Kingdom

*Abstract:* Flow induced orientation, the subject of this article, is being surveyed in its wide variety of manifestations. This includes crystallisation together with the resulting orientation in the solidified product, a broad survey of past works here placed in a new perspective not presented hitherto, and novel rheological effects observed lately in melt flow containing some recent results, some to be announced here for the first time. The theme connecting the two separate subject areas is the coil → stretch transition produced by the appropriate flows allowing for only two states, the essentially random coil and the essentially fully stretched out state, with no intermediate state under steady state conditions in between. Each of the above two states leads to its own characteristic morphology on crystallisation which then characterise the final product and the resulting orientation pattern as registered by x-ray diffraction. It is being argued that the whole range of orientation patterns observed in articles crystallised under melt flow are accountable by the variable ratios of the two crystal morphologies, lamellae and fibres arising from the respective random and fully aligned chain orientations, through their various combinations, without the need for involving intermediate chain or crystal orientations at least as relating to the strength of the primary orienting influence. — The novel rheological effects are arising from the main theme, namely the random and extended chain duality, with criticalities in terms of both strain rate and molecular weight sharply delineating the two states. This, when coupled with a so far unspecified phase transition, can provide an unexpectedly sharp temperature singularity (window) in extrusion behaviour potentially highly advantageous for processing applications. With constant piston velocity this is manifest as a sharp minimum in extrusion pressure, and at fixed extrusion pressure as a sharp maximum in extrudate exit velocity, the criticality in either case displaying an inverse fourth power dependence on molecular weight. — In order to provide an underpinning for both classes of effect raised above, a brief survey is provided on the fundamentals of elongational flow induced coil → stretch transitions as explored experimentally in the Bristol laboratory on polymer solutions. This is accompanied by a discussion of the transferability of those results to melts including both the parallels between the two systems and the apparent problems such a transfer seems to raise, ending on some open ended issues as regards similarities and differences between freely flowing melts and extension of connected networks.

*Key words:* Chain extension — orientation — free flowing melts — extension of connected networks

## Introduction

It hardly needs stating these days that long chain molecules lend themselves to the attainment of oriented products with associated advantageous properties. There are principally two different routes towards this goal: 1) drawing (or otherwise orienting) an initially random crystalline solid, and

[1]) Permanent address: DSM Research BV, Koestraat, P.O. Box 18, 6160 MD Geleen, The Netherlands.

2) to orient chains in their random state (solution or melt) first and "set" this orientation by subsequent crystallisation. In this article we shall be concerned with the second route only. We are aware that this has been done repeatedly before by one of us [1—3]. Yet, in the present paper the pertinent material will be presented with a specific underlying theme, to our knowledge not previously featured, at least in the explicit manner as it will here. In aid of the same we shall invoke some work on the fundamentals of chain extension, performed on solutions. Again the latter has been presented before, but this in the context of solution behaviour; its reinvoking in connection with crystallisation may therefore not be an inopportune repetition in view of the context of this conference. Finally, we shall quote from some current work on melt rheology, partly because of its intrinsic interest and partly because in our opinion it underpins the main theme we are attempting to transmit in this paper.

The theme is as follows: in terms of chains there are only two stages of orientation (within subject area 2) above): the essentially fully stretched out and the essentially unoriented random chain, with no intermediate stages which are of consequence for us here, and further, that all we observe in the final products are the consequences of these two extremes. Intermediate orientations, such as are seen between in partially, or not fully oriented products, are due to the different combinations of the two extremes just stated, and not to intermediate chain orientations as such. We consider these assertions as basic with a vast and diverse amount of observational evidence and also some fundamental background in support of it, which here we shall attempt to convey.

There are two stages to be considered: i) The primary chain orientation itself. ii) Fixation of the chain orientation, which occurs mainly by crystallisation. Our theme of two extremes, with no intermediate stage, applies to both stages but in different interrelated contexts. Thus when a single chain within an assembly which is mechanically unconnected to others, is exposed to external orienting influences (in our case flow), it either extends practically completely or stays a random coil. This of course is a bold generalising statement not intended to imply that departures from the fully stretched towards the contracted state do not exist. Clearly they do, but we maintain that within steady state conditions such departures are small, and as a first approximation do not invalidate our main theme. We shall "lump" such departures together within the random or fully oriented categories according to which they are closest. Of course there must be intermediate states as transients (as opposed to the steady state) on the way to full chain extension, but it happens that in the wide ranging effects to be referred to here we have yet to note their influence; so far we found no need to take note of them.

In the above we were referring to the primary chain extension. When passing over to crystallisation we again we have two distinct categories arising from the above two extremes: the random chain leads to chain folded lamellae and the extended chain to fibrous crystals, with the chains essentially (including all kinds of defects) extended within them. These two morphologies are again distinct, with to our knowledge, nothing intermediate in between.

The question which remains is the degree of chain extension which is needed to "switch" from the lamellar to the fibrillar morphology. In situations where the chain orientation prior to crystallisation can be diagnosed (such as in solutions, see below) the question does not arise because no intermediate state of chain extension, no "half way house", could be found, at least as a steady state. In cases where a similar diagnosis could not be carried out with the same conclusiveness (as e.g. in melts) the question remains as to the degree of chain extension required for the morphology to "switch" from lamellae to fibres within a continuous spectrum of chain extensions. Nevertheless, as will be seen, the situation as regards "on-off" (or "yes-no") effects so closely parallels the solution case that a direct transfer of experience from solutions to melts is clearly invited: more specifically, the existence of a coil → stretch transition and the exclusive (to a first approximation) duality of the random and fully extended chains. This is in spite of difficulties to envisage such a situation in the molten state, be it for intuitive or preconceived reasons.

In what follows we shall try to combine works on flow induced crystal morphology with those on flow induced chain extension, the latter irrespective of crystallisation or preceding crystallisation. Historically, at least in our hands at Bristol, the elucidation of the crystallisation effects was the first objective which then, for deepened understanding, prompted a separate program on chain extension itself. The latter line pursued for the simpler case of solutions, has become a pursuit of its own, providing, so we hope, an improved insight into the extensibility of long chain molecules. The present article is to bring these two subject areas, stemming from the same roots, but pursued separately since, together again in service of the theme stated. The main point is that there is a twoway connection between chain extension and crystal morphology resulting therefrom: the chain extension determines the morphology on subsequent crystallisation, while the morphology, as observed after crystallisation, can serve as a pointer to the preexisting state of chain extension, provided we can read the message it conveys.

## The fundamentals of chain extension as explored on solutions

### *Preliminaries and methodologies*

First the fundamentals of chain extension will be recapitulated as emerging from our works on solutions. As will be familiar, extending a long chain substantially is not as straightforward as it may first appear. For uncrosslinked chains stretching forces can only be applied by means of flow through frictional contact between solvent and chain (in case of solutions). The customary simple shear flow is inadequate in this respect because of its rotational component. To achieve full, or nearly full, chain extension the flow has to be "elongational", i.e. the dominant component of the velocity gradient has to be parallel to the flow direction, the simplest case of the latter being unidirectional accelerating flow. The effect of elongational flow on an isolated random coil macromolecule, as to be found in dilute solutions, was worked out theoretically by De Gennes [4] (following preceding works by Peterlin [5]) leading to the prediction of a sharp coil → stretch transition at a specific strain rate ($\dot{\varepsilon}$). This means that as the elongational flow rate is being gradually increased through increasing the velocity gradient along the flow direction, there will first only be small changes in the conformation of the random coil up to a critical strain rate ($\dot{\varepsilon}_c$) beyond which the coil will stretch out practically fully. It follows that there is no steady state intermediate state: below $\dot{\varepsilon}_c$ there is practically no chain extension, while above $\dot{\varepsilon}_c$ the chain becomes practically fully extended. In addition to reaching the critical strain rate sufficient strain ($\varepsilon$) is also required namely, that the extensional influence is maintained for a long enough time for the final steady state extension to be reached. The latter can be achieved through appropriate flow field geometry. Thus if there is a stagnation point present the strain condition will automatically be satisfied.

In plain words, the origin of the coil → stretch transition lies in the fact that once the resistance of the random coil to substantial extension is overcome on reaching $\dot{\varepsilon}_c$, its frictional contact with the surrounding fluid (solvent) medium is being increased, which in turn increases the resulting chain extension and so on. Thus the extension of the chain becomes a run-away effect until near to complete chain extension further extension is opposed by the intrinsic limit to the stretchability of the valence bonds themselves.

It is apparent that the above considerations are of salient significance for the response of macromolecules to flow, and vice versa, for the effect of macromolecules on the flow itself. This is the subject area which the Bristol program set out to explore during the last decade [6—14] of which some landmarks will be recalled in what follows.

The first requirement was the creation of well characterised elongational flow fields, and secondly a diagnostic tool to follow chain extension when this occurred. For the former several methods were designed and/or used. Foremost amongst them was the flow cell consisting of two opposed jets. The two jets were immersed in the solution which was being sucked out simultaneously through both jet orifices. This created a purely elongational flow field between the two jet orifices along the centre line. Here the velocity is exactly zero at the centre (stagnation point) and increases linearly in both directions towards the jet orifices (Fig. 1). Chain orientation was registered by observing the resulting birefringence between crossed polaroids. This became apparent as a sharply defined bright line against a dark background along the central jet axis [10] (Fig. 2). When monodisperse (sharp molecular weight ($M$) distribution) polymer was used this bright line appeared discontinuously at a critical strain rate ($\dot{\varepsilon}_c$) as $\dot{\varepsilon}$ was being increased. The degree of chain stretching could be determined from the magnitude of the birefringence which, within the latitude of known polarisability values was found to correspond to closely full chain extension [6]. These findings thus confirmed the reality of a coil → stretch transition in elongational flow in support of theory; conversely it established the method, possibly the only method, by which a

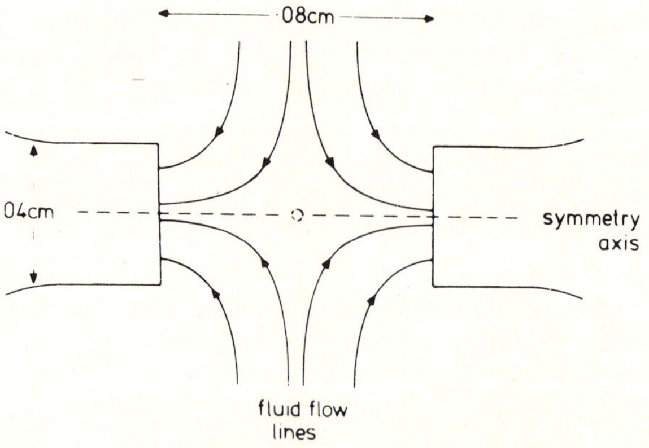


Fig. 1. The principle of the opposed jet technique for the study of molecular extension by elongational flow showing flow lines

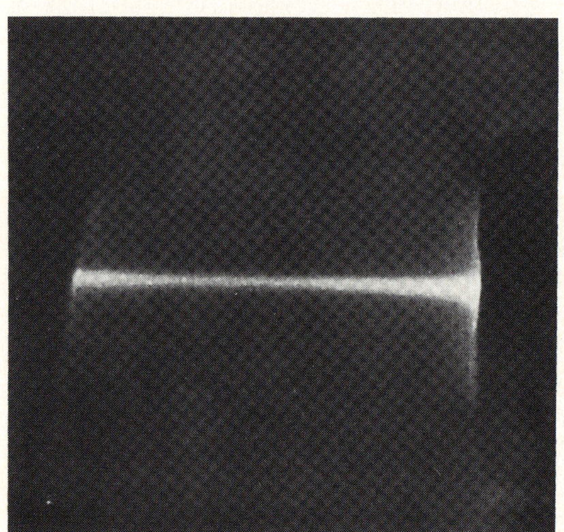

Fig. 2. Chain extension by elongational flow between two opposed jets seen as a birefringent line between crossed polars, in this case in a polyethylene solution above the crystallisation temperature [10]

long-chain flexible molecule can be stretched out nearly fully in a controlled and registrable way. The sharpness of the lines could be satisfactorily attributed to the fact that they corresponded to flow paths going through or near the stagnation point where the deforming fluid element spent longest time, hence where for a given strain rate ($\dot{\varepsilon} \geqslant \dot{\varepsilon}_c$) the required strain ($\varepsilon = \dot{\varepsilon}t$) for full chain extension could be achieved. It will become apparent that we acquired a powerful visual method for probing the influence of flow on macromolecules and vice versa, also the influence of macromolecules on the flow itself, which has been exploited in works which followed.

*Molecular characterisation by flow methodology; flow induced scission*

One of the follow ups of the above mentioned recognitions was the relation between the critical strain rate ($\dot{\varepsilon}_c$) and the molecular weight ($M$). First the basic quantitative aspect of the experiment will be recalled. For this the transmitted light intensity, $I$ (which for low optical retardations is proportional to the square of the retardation), along the central axis of the jet is plotted as a function of $\dot{\varepsilon}$. As seen, say from the curve for $M_1$ in Fig. 3, $I$ is practically zero till $(\dot{\varepsilon}_c)_1$ when it suddenly rises to a plateau value. Here $(\dot{\varepsilon}_c)_1$ is the critical $\dot{\varepsilon}$ of the coil $\rightarrow$ stretch transition for material of molecular weight $M_1$ and the plateau value of $I$ corresponds to that due to the practically fully stretched out chain. Now, taking closely monodisperse polymers of different $M$ ($M_2$ and $M_3$ in Fig. 3), we obtain different

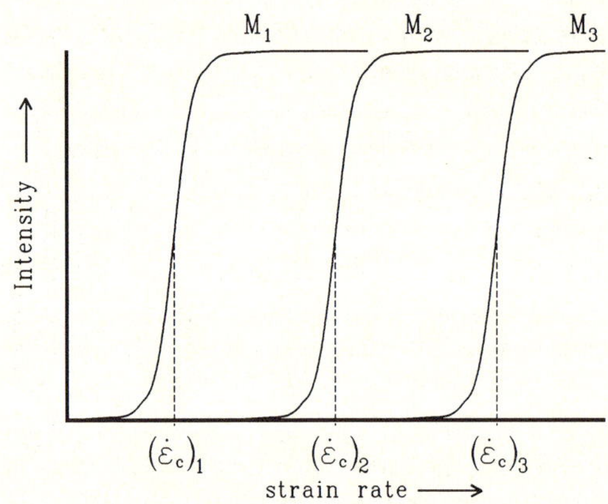


Fig. 3. Schematic plot of transmitted intensity vs. strain rate for three different molecular weights $M_1 > M_2 > M_3$, showing the coil $\rightarrow$ stretch transition at a critical strain rate, resp. $(\dot{\varepsilon}_c)_1$, $(\dot{\varepsilon}_c)_2$ and $(\dot{\varepsilon}_c)_3$

curves with different $\dot{\varepsilon}_c$ values. The schematics in Fig. 3 illustrate the trend: the lower $M$ the higher $\dot{\varepsilon}_c$ and vice versa. This means that longer chains can be extended by lower strain rates, hence less strong flows, and vice versa: in plain words the long chains are more readily extensible. Quantitatively this is expressed by the relation:

$$\dot{\varepsilon}_c \propto M^{-\beta} , \qquad (1)$$

where $\beta$ was found to be 1.5 by our experiments, rather remarkably, independent of solvent quality [7].

It will be readily apparent that the newly found relation Eq. (1) and the method to determine $\dot{\varepsilon}_c$ experimentally provides a new route towards determination of molecular weights with certain advantages over conventional methods [9, 11]. First, it is most sensitive at the highest $M$ (where $\dot{\varepsilon}_c$ is lowest), where most other methods are either too insensitive or are difficult to apply. Secondly, when there is a continuous distribution of molecular weights (as is usually the case) then the resultant $I$ vs. $\dot{\varepsilon}$ trace will be a sum of components such as $M_1$, $M_2$, $M_3$ etc. in Fig. 3 which can be readily dissected into its components by differentiation. This then provides a method for obtaining the actual molecular weight distribution from the highest $M$ end to a low cut-off which by Eq. (1) is determined by the highest $\dot{\varepsilon}$ that can be achieved in a given apparatus. Thus we see that a flow induced effect can provide a tool for basic molecular weight characterisation by utilising chain extensibility, possibly the most intrinsic characteristics of a flexible filamentous molecule.

We thus see that the coil → stretch transition, as assessed by the single parameter $\dot{\varepsilon}_c$, has a two-fold critically: for a given $M$ it is critical in $\dot{\varepsilon}$ and for a given $\dot{\varepsilon}$ it is critical in $M$. The latter acquires particular significance in the usual situation of a molecular weight distribution when subjected to an elongational flow field of a particular $\dot{\varepsilon}$. Here the longest chains only will be extended with a low end cut-off at a specific $M$ corresponding to the appropriate $\dot{\varepsilon}_c$ by Eq. (1) while the rest of the molecules will remain unstretched. Thus we shall have a bimodal distribution of chain extension constituted by practically fully stretched out and essentially unstretched chains with no intermediate stage in between, a result of salient importance for what follows. Increasing $\dot{\varepsilon}$ will not substantially increase the degree of chain extension (i.e. in terms of end-to-end distance), but will increase the amount of material which becomes extended within a bimodal distribution of extended and unextended chains by increasingly "cutting into" the distribution from the high molecular tail downwards. This situation is illustrated by Fig. 4 with reference to three strain rates ($\dot{\varepsilon}_c$) of increasing magnitude with the three $M$'s in Fig. 3.

We now turn to a further notable aspect of the present type of chain stretching experimentation in

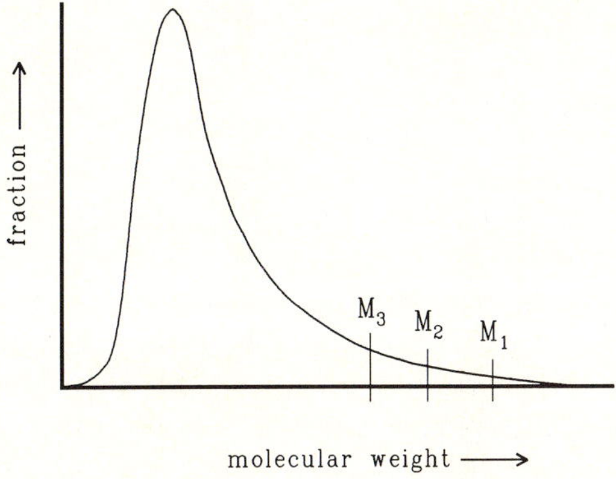


Fig. 4. Schematic plot of molecular weight distribution showing the three different cut-offs for molecular weights corresponding to Fig. 3. (For explanation see text)

elongation flow fields, namely flow induced chain fracture. Consider a given molecular weight, say $M_1$, in Fig. 3 and continue to increase $\dot{\varepsilon}$ beyond $(\dot{\varepsilon}_c)_1$, i.e. proceed along the plateau of the $I$ vs. $\dot{\varepsilon}$ curve. Here, at some stage, the fully stretched out chain will break as a result of increasing stretching forces along its length transmitted by frictional contact from solvent to chain. It is the salient result of our experiment that the chains were shown to break exactly at the centre. The latter was established by using the same experimental set up by which the chains had been broken initially for determining the molecular weight of the scission fragments by our elongational flow method as described in the previous paragraph (For a recent review of Bristol works on chain scission, see ref. [8]). This chain halving through flow induced scission, exact within the accuracy of available methods for molecular weight determinations, is perfectly accountable theoretically and is presently one of the corner stones of our understanding of mechanically induced chain degradation in flowing systems. We shall invoke it again in support of our interpretations of the novel melt rheological effect in section "Implications" (see footnote there).

*Extension of interacting chains; creation of entanglements*

All the above sofar related to the simple isolated chain as it exists in these experiments on dilute solutions. In more concentrated solutions (and of course in the melt) the chains overlap and are entangled. Performing the same experiments using opposing jets had important information to convey on the effect of flow on the entangled systems. In such systems two stages were found on increasing the strain rate, more explicitly two critical $\dot{\varepsilon}$'s. The first corresponds to the coil → stretch transition as before. That is as $\dot{\varepsilon}$ is increased, at some $\dot{\varepsilon}$ a bright central line appears between the jets as in Fig. 2 signalling $\dot{\varepsilon}_c$, hence a coil → stretch transition. Increasing $\dot{\varepsilon}$ further, with constantly new solution containing unstretched molecules entering the system, a second distinct event sets in (following some intermediate stages not to be enlarged upon here). Instead of the steady central bright line the birefringence pervades the whole volume between the jets, the system becoming unsteady, the birefringence flickering like a flame, a still picture of which is being given by Fig. 5. This we consider as the manifestation of the formation of a mechanically connected network [12, 13]. At the strain rate $\dot{\varepsilon}_n$ when the "flare" such as in Fig. 5 sets in, the chains do not extend individually under the influence of flow, hence the flow becomes unstable.

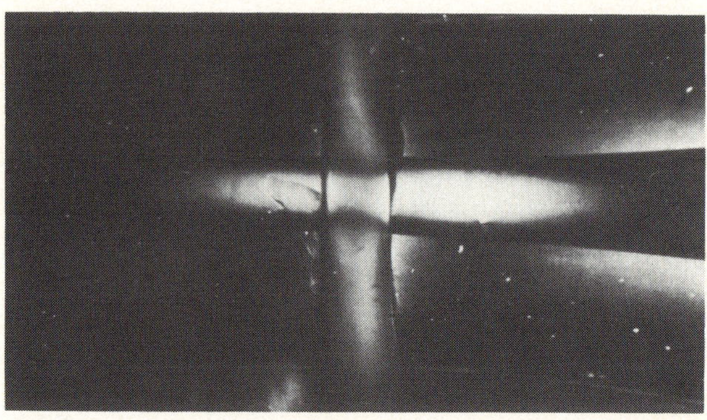

Fig. 5. Optical micrograph observed between crossed polars showing the "flare" phenomenon. 0.1% a-PS solution. $M = 7.2 \times 10^6$ [12]. (For explanation see text)

There are two criticalities defining these happenings: 1) a critical polymer concentration, $c^+$, at which (and beyond) the above "flare", signalling network formation, sets in, and 2) for a solution of concentration above $c^+$, a critical strain rate $\dot{\varepsilon}_n$ needs to be reached where $\dot{\varepsilon}_n > \dot{\varepsilon}_c$ [2]).

[2]) $c^+$ is substantially smaller than the conventionally defined concentration for chain overlap i.e. $c^*$. The reason is that the elongational flow induced interchain interaction requires a lesser degree of chain overlap than corresponds to contact at radius of gyration, the criterion used to define $c^*$.

Figure 6 attempts to convey what is happeing. It depicts an overlapping system of chains. As the strain rate reaches $\dot{\varepsilon}_c$ the chains disentangle and stretch out individually (Fig. 6a). As $\dot{\varepsilon}$ is increased, while always fresh solution with overlapping molecules is exposed to it, a second critical

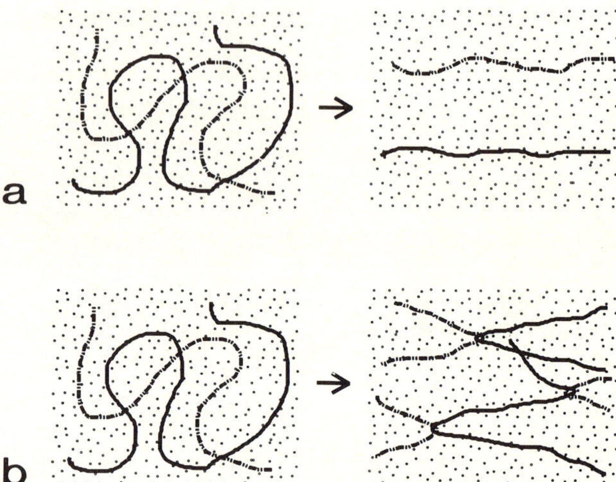


Fig. 6. Schematic drawing of the respons of dissolved molecules to an elongational flow field (at $c > c^+$). a) $\dot{\varepsilon}_c < \dot{\varepsilon} < \dot{\varepsilon}_n$ (within strain rate window). b) $\dot{\varepsilon}_c < \dot{\varepsilon}_n < \dot{\varepsilon}$ (mechanically effective network formation giving rise to the "flare" phenomenon)

strain rate $\dot{\varepsilon}_n$ is reached, when the chains, now on the much shorter time scale corresponding to $\dot{\varepsilon}_n$ "grip each other", so to speak, and will act like a mechanically connected network (Fig. 6b). By separate work it was established that at this stage also the flow resistance (viscosity) is enormously increased and flow induced chain scission, which now becomes increasingly random, is greatly enhanced [13, 14]. We see therefore that by simple visual criteria such as Figs. 2 and 5, we can explore the effect of flow on molecules, use these effects to characterise them and diagnose interactions between them.

The relation between $\dot{\varepsilon}_c$, $\dot{\varepsilon}_n$ and concentration ($c$) is expressible in graphs such as Fig. 7. This is a schematisation of numerous results reported in specific detail in the individual works. All these have the following features in common: i) $\dot{\varepsilon}_c$ is a slowly decreasing function of $c$. This is the consequence of the increasing *solution* viscosity with increasing $c$ which in turn increases the conformational relaxation time $\tau = 1/\dot{\varepsilon}_c$ (see refs. [5, 12—14].

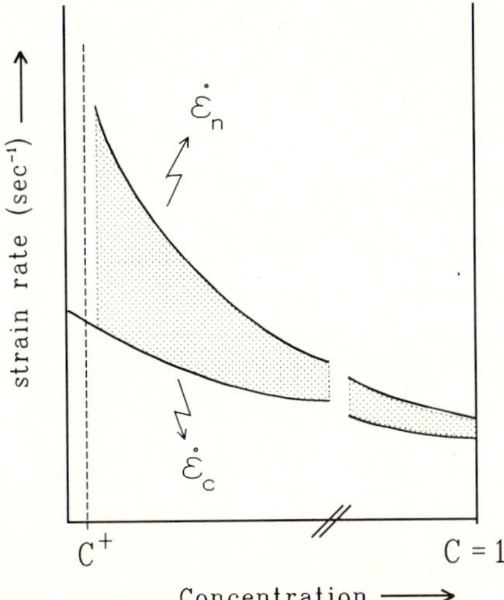


Fig. 7. Strain rate vs. concentration plot, showing the critical strain rate for the coil → stretch transition ($\dot{\varepsilon}_c$) and the critical strain rate for network formation ($\dot{\varepsilon}_n$) in case a finite strain rate window ($\dot{\varepsilon}_n$ — $\dot{\varepsilon}_c$, shaded area) pertains up to $c$ = 1 (single component melt)

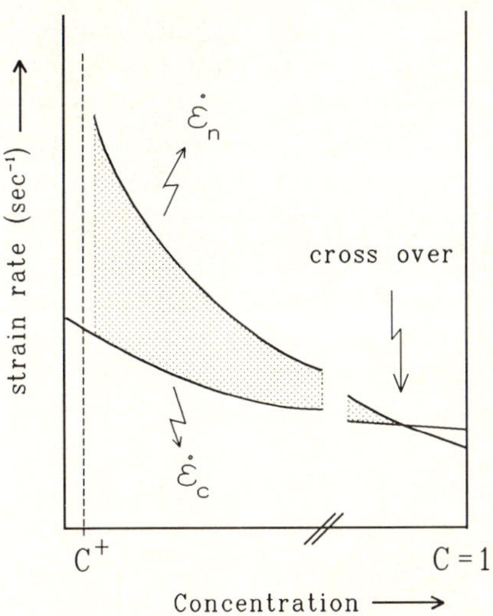


Fig. 8. Strain rate vs. concentration plot as in Fig. 7, but now for the case of a cross over of the $\dot{\varepsilon}_n$ and $\dot{\varepsilon}_c$ vs. concentration lines, in which case the strain rate window ceases to pertain with increasing concentration

It is important to note that the extending chain responds to the frictional forces corresponding to the mean viscosity of the medium surrounding it (i.e. the viscosity compounded by solvent and solute). ii) $\dot{\varepsilon}_n$ (where $\dot{\varepsilon}_n > \dot{\varepsilon}_c$) becomes operative, i.e. the flare sets in — Fig. 5, at and beyond a particular $c \equiv c^+$. This means that for $c \geqslant c^+$ there is an upper bound in $\dot{\varepsilon}$ where the chain is extensible individually, hence there exists a strain rate "window" $\dot{\varepsilon}_n$ — $\dot{\varepsilon}_c$ (shaded area in Figs. 7 and 8). iii) As $\dot{\varepsilon}_n$ is found to decrease more steeply than $\dot{\varepsilon}_c$ the strain rate window $\dot{\varepsilon}_n$ — $\dot{\varepsilon}_c$ narrows with increasing $c$. iv) $\dot{\varepsilon}_n$ seems to approach $\dot{\varepsilon}_c$ asymptotically, with $\dot{\varepsilon}_n$ — $\dot{\varepsilon}_c$ narrowing accordingly, but with no evidence of a cross-over of $\dot{\varepsilon}_n$ with $\dot{\varepsilon}_c$ as a function of $c$. This is highly sifnificant as the existence of a cross-over (Fig. 8) would signify that a given molecule cannot be extended any longer in its entirety, because it would become portion of a mechanically active network before that, as $\dot{\varepsilon}$ is being increased. It follows that as long as condition iv) holds a chain will remain extensible within its appropriately narrowed down $\dot{\varepsilon}_n$ — $\dot{\varepsilon}_c$ strain rate window at the correspondingly lowered strain rate.

The question which arises at this point is how far the condition in Fig. 7, i.e. the existence of a finite strain rate window of chain extensibility, holds with increasing $c$, and specifically, wether it still holds for $c$ = 1, i.e. for the single component melt. As will be seen in the final section the whole body of evidence in our work on themelt is consistent with the conditions embodied by Fig. 7, i.e. that the extensibility of the full chain pertains even there. Even so, situations will be quoted (cross-linked melts) where this cannot be so a priori and where $\dot{\varepsilon}_n < \dot{\varepsilon}_c$ must pertain, i.e. where we must be at the high $c$ side of a crossover such as in Fig. 8, but where nevertheless morphological effects equally attributable to the fully extended chain (but now between fixed entanglement points) can be seen. The existence or non-existence of a crossing over of $\dot{\varepsilon}_n$ with $\dot{\varepsilon}_c$, and its dependence (if and when it exists) on $c$, and on preexisting cross-links, is clearly crucial for the attainment of full chain extension. It will therefore be crucial also and beyond, for the understanding of entanglements and melt behaviour in general and will remain the open ended issue in this paper.

## Coil → stretch transition and fibre-plateled duality

### *Shish-kebab crystals from solution*

The most conspicuous consequence of the "on-off" effect of chain stretching by elongational flow is the formation of shish-kebabs as first systematically

Fig. 9. Electron micrograph showing the shish-kebab morphology as obtained from solution [45]

produced and investigated through studies of stirred solutions, largely initiated by the pioneering works of Pennings [15], of which Fig. 9 should serve as a reminder. Historically, the emphasis was first on the accounting for the dual fibre-platelet morphology. As known this was done through visualising the coexistence of highly stretched and essentially unstretched chains where, the former crystallise first giving rise to fibres of which the unstretched chains then avail themselves as nucleating templates for their deposition in the form of chain folded lamellae.

For our present purpose we are now using this original argument and are invoking the ubiquitously documented shish-kebabs as evidence for the preexisting duality of extended and unextended chains prior to crystallisation through our basic studies on elongational flow induced chain extension in closely monodisperse material of noncrystallisable polymers. In Pennings' original works the source of elongational flow was identified as convergent and divergent flow field regions between Taylor vortices. In the subsequent studies by ourselves on elongational flow induced chain extension such flow fields were created purposefully in the approximately constructed apparatus (jets, slots, rollers etc.). Figure 10 should provide the link between the two lines of work where, using the crystallisable polymer polyethylene, flow induced chain extension was created and registered first and fibrous crystallisation was induced subsequently at the appropriately lower temperature [3, 10]. Electron microscopic examination of fibres, such as seen lightoptically in Fig. 10, indeed confirmed the shish-kebab character of the underlying crystals.

Investigations in the opposed jet apparatus, such as underlie Fig. 10, have been taken up more recently again, in the light of knowledge acquired

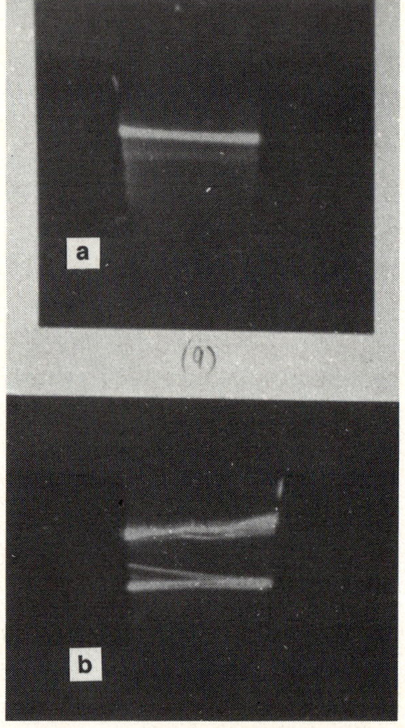


Fig. 10. a) The development of crystallisation observed during suction in the opposed jet apparatus for a 3 wt% polyethylene solution at $T = 100\,°C$, $\dot{\varepsilon} = 2.3 \times 10^3\ s^{-1}$, $M \approx 200\,000$. Polars crossed at 45°. b) Crystal deposit obtained during suction which has blocked one of the jets. Polars crossed at 45° [3]

since. Specifically, it addressed the question as to whether or not shish-kebabs can arise from initially nonoverlapping chains when suitably extended or whether chain overlap, i.e. network formation, is necessary for their formation [16]. This distinction has become possible since the recognition of the flare (Fig. 5) as a criterion for chain overlap. For this purpose experiments were performed on polyethylene as a function of polymer concentration testing for $\dot{\varepsilon}_c$ and $\dot{\varepsilon}_n$, as laid out above by the sharp line and flare criteria respectively (Figs. 2 and 5), first at a temperature sufficiently high to avoid crystallisation and thus establish the value for $c^+$. Subsequently fibres were produced by performing the same experiment at appropriately lower temperatures. It was found that fibres with shish-kebab character did form, both above and below $c^+$. We quote these unpublished results at this place in view of recent reports from elsewhere [17] that gels (hence networks) are necessarily involved in shish-kebab formation. While not denying that stretching of entangled systems, such as gels, gives rise to shish-kebabs (an important point to emerge further below), by evidence just quoted this need not be a necessary condition, as according to it shish-kebab crystals can arise also from stretched out unentangled chains.

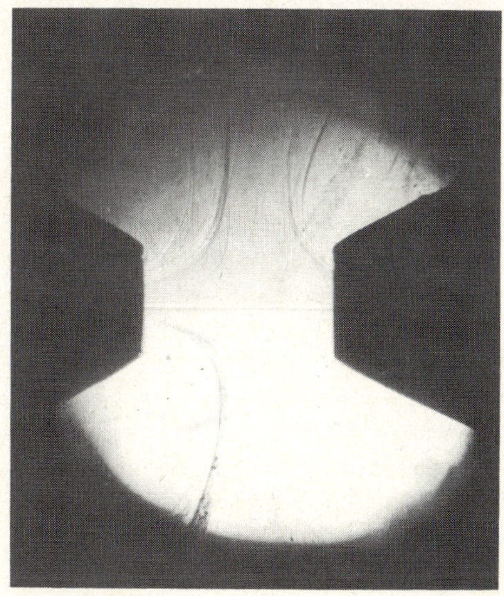

Fig. 11. Bright field photograph taken during flow of region between orifices of the opposed jets apparatus showing formation of central fibre and additional fibrous crystallisation induced by placing a gauze upstream in the polymer melt supply. Polyethylene, $T$ = 140°C [18]

*Elongational flow generated fibrous crystals from the melt*

As stated in the introduction we cannot diagnose chain extension in the melt with the same definiteness as we can in solutions. One reason is the lack of similarly well defined crystallisable material in sufficient quantities needed for such experiments: the high molecular weight polyethylene available for such a purpose is inevitably highly polydisperse. Nevertheless, the recalling of examples of early jet experiments, using PE melts [3, 18, 19] can be instructive. In Fig. 11 clearly defined fibres appear along the central line between the jets, seen most clearly in unpolarised light. The resemblance to the corresponding solution case, both before and after crystallisation, is evident (compare Figs. 10 and 11). By this, to us compelling evidence it thus appears that given the appropriate strain rate and strain (ensured by the stagnation point along the centre axis) molecules in the melt can align and stretch out just as they do in solutions, which in Fig. 11 is made visible by the fixation of the stretched out state through crystallisation. Another feature of Fig. 11 deserves pointing out: namely the fibres along parabolic stream lines entering the jets from

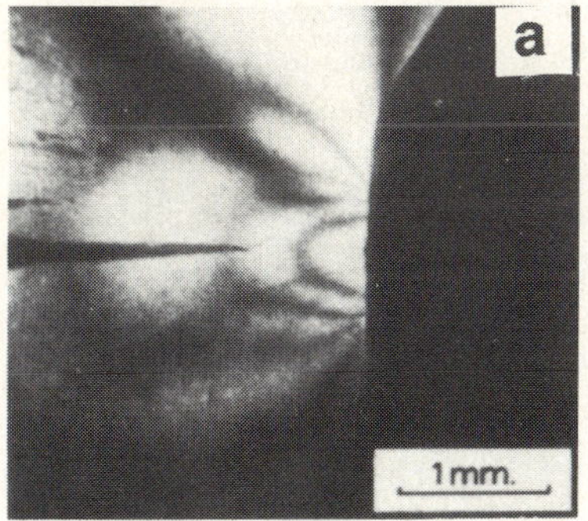


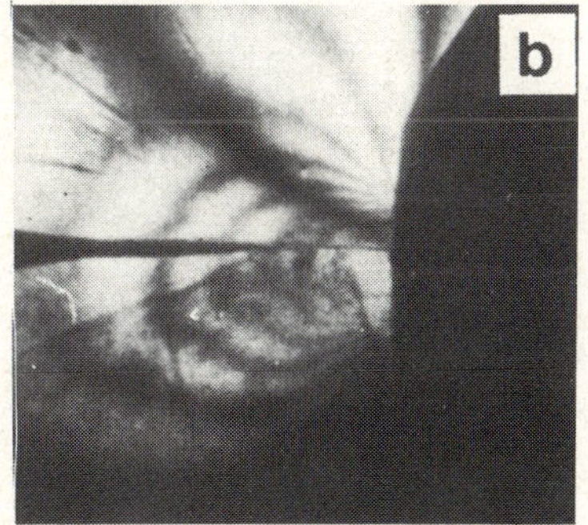


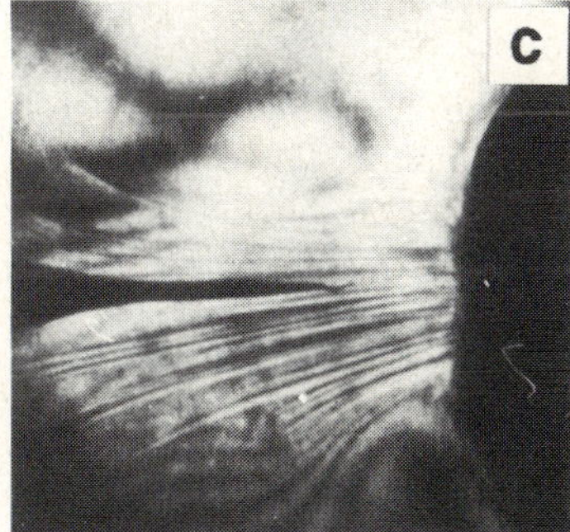


Fig. 12. Photographs viewed between crossed polars showing the development of fibrous crystallisation for a polyethylene melt flowing into an orifice with a stationary needle-like obstruction upstream. $T$ = 140°C. a) Mean velocity into orifice ($v_m$) = 0.8 cm s$^{-1}$, b) $v_m$ = 3.3 cm s$^{-1}$, c) $v_m$ = 8.0 cm s$^{-1}$ [19]

the top downwards. These, as recognised at the time [18], originate behind the bars of a metal grid placed there as a filter, which in turn generate stagnation points for flow downstream like any stationary obstacle would do as a matter of course. The last point is brought out most strikingly by Fig. 12b where a fibre is seen to originate downstream from the tip of a needle (hence from the stagnation point there arising) deliberately placed there for that purpose.

In Fig. 12c, on increasing the strain rate, a further effect is seen, namely fibres entering the single jet along the converging flow lines across the full jet diameter. This corresponds to the generation of fibres themselves due to underlying chain extension, by a single orifice where (as opposed to the double jet) there is no stagnation point, at least as defined by the single jet geometry on its own. It follows that for the same overall strain rate the achievable strain here will be smaller than behind the needle tip at the locality of the stagnation point. Indeed, as seen from Fig. 12c, the strain rate had to be increased (as compared with Fig. 12b) for this type of fibre formation in a single jet to set in. Fibre formation in flowing melts entering a single jet has been treated separately in an earlier publication from this laboratory [19], where, amongst others, conditions of a self stabilising dynamic equilibrium resulting from the interplay between the velocity gradient of the overall flow field and the velocity gradient locally arising in the course of the growing fibre is being defined and analyzed. For particulars we have to refer to that publication. Here the single jet case and its role in melt flow is being invoked merely because it provides a link with the work relying on a capillary rheometer to be described in the sections to follow. Clearly, the situation with the single jet in ref. [19], from which Fig. 13 has been taken, closely resembles that of the capillary entry in the rheometer, the subject of more recent studies both on structure and the rheology from which we shall quote in what follows.

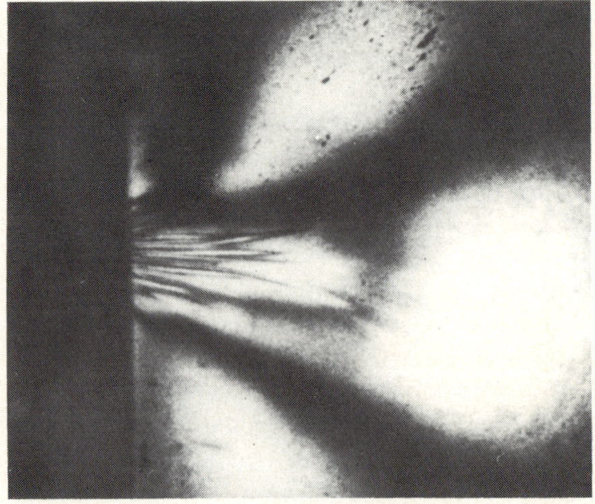

Fig. 13. Photograph viewed between crossed polars showing the development of crystallisation. As in Fig. 12 but here without obstacle. $v_m$ = 2.6 cm $s^{-1}$ [19]

*X-ray orientation effects in terms of fibre-platelet duality*

For what follows we revert again to our initial theme, namely that in orientation induced by flow we have essentially two types of entity: those which are practically fully extended and those which are practically unextended, with (in first approximation) nothing in between. The "entity" can refer to two different dimensional levels, namely to the chains and to the crystals arising therefrom. Subsequently we gave attention to the chain, and in what followed we carried this over directly to the crystalline state on the basis of causal arguments (shish-kebabs in solution) and visual resemblance (melts). In what follows we shall focus on the various manifestations of flow induced crystallisation from the melt upholding the main theme. First, this will be done descriptively to which then a few comments will be added on possible connections with the preceding molecular considerations, which as will be seen, will lead up to some hitherto unrecognised open ended issues.

The common ground to it all is that flow of the appropriate kind and strength will extend the chains in a limited portion of the material giving rise to fibrous crystals aligned along the flow field with the remainder of the material crystallising into chain folded platelets just as it would do in a state of rest. Thus, the flow oriented final crystalline product is constituted by the combination of the above two morphological entities. In what follows we shall look at the possible combinations. In the first place this will be irrespective of how they originate on a molecular level, merely restating the fact that, in first approximation, increasing the strength of the flow creates more fibres, i.e. it orients and extends more chains, as opposed to extending all the chains to a higher degree.

At the high flow strength end of the spectrum a situation when all the material is converted into fibres is being at least approached. This leads to the familar situation of blockage in the capillary rheometer, correctly attributed by van der Vegt [20] to elongational flow induced chain stretching generated crystallisation. As known from numerous studies by Porter and coworkers [21] the solidified plugs arising in this way consist largely of extended chain type fibrous crystals.

At the other end of the spectrum when the flow is very weak only very few fibres form. These, in themselves, would be inadequate to influence the final sample characteristics if it were not for setting the pattern of the ensueing lamellar overgrowth for the overwhelming major portion of the material, which has remained unextended. This will impart a characteristic texture, on the morphological level, and a characteristic orientation pattern, on the level of x-ray diffraction diagnosis, to the sample as a whole, to be taken in turn in what follows.

On the morphological level the characteristic effect is a fibrosity along two mutually orthogonal directions, one parallel and the other perpendicular to the initial orienting influence (flow), but on different scales. At low magnification it will be the parallel fibrosity which is apparent, represented by the original flow-induced fibrils plus the surrounding envelope of transverse lamellae, where the latter may not be resolvable individually. The impression of parallel fibrosity is thus being created by the parallel columns consisting of core fibres and overgrowth platelets. At high magnifications, on the other hand, the transversely growing lamellae will appear prominent giving the overall impression of a fibrosity which is perpendicular to the flow direction, seemingly contrary to a priori expectations. Clearly the borderline between the two cases, as perceived at a given magnification, will be much affected by the width of the columns, themselves determined by the number of fibres, in turn determined by the strength of the flow.

Regards crystallographic orientation the fibres will always impart a *c* axis orientation parallel to the flow direction. As long as the transverse lamellae are all parallel to each other (Fig. 14c), they too will represent *c* axis orientation w.r. to the flow direction imparting the usual fibre orientation with the familiar *c* axis diffraction pattern (Fig. 15a and sketch 14d) to the sample as a whole. (This, amongst others, is the case for a parallel assembly of solution grown shish-kebabs). In such a structure the fibre and platelet components are indistinguishable by the usual wide angle x-ray diffraction pattern. However, in situations when fibres are few, hence widely spaced, as arises in weak flow, then the transverse lamellae will twist (hence randomise around *b* in case of polyethylene) as they do in a spherulite but here confined to planes perpendicular to the flow direction (Fig. 14a). This type of arrangement, the originally termed "row structure", gives rise to a characteristic x-ray diffraction pattern (Fig. 16) sketched in Fig. 14b readily derivable by pole figure constructions from a structure such as Fig. 14 [22, 23]). To note, such a diffraction pattern, commonly displayed by the usual hauled off or

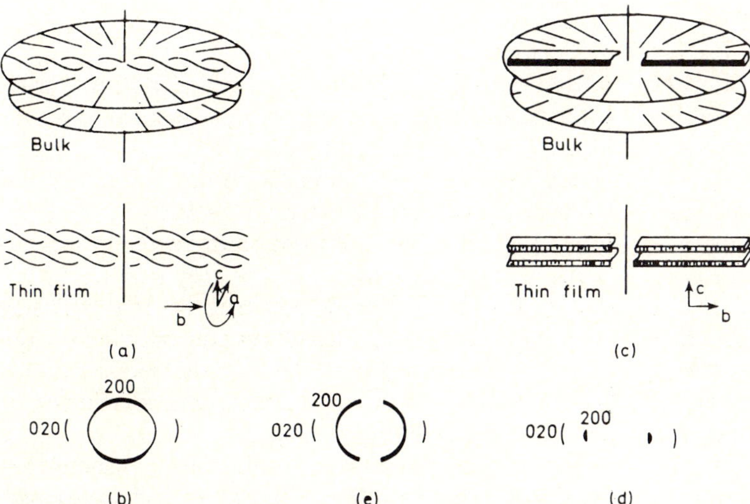


Fig. 14. Schematic representation of the crystal texture originating during the crystallisation of oriented polymeric melts shown for the particular case of polyethylene. a) Low stress. b) The main features of the x-ray diffraction pattern corresponding to a). c) High stress. d) The main features of the x-ray diffraction pattern corresponding to c). e) The main features of the x-ray diffraction pattern for stresses intermediate between those in b) and d) [3]

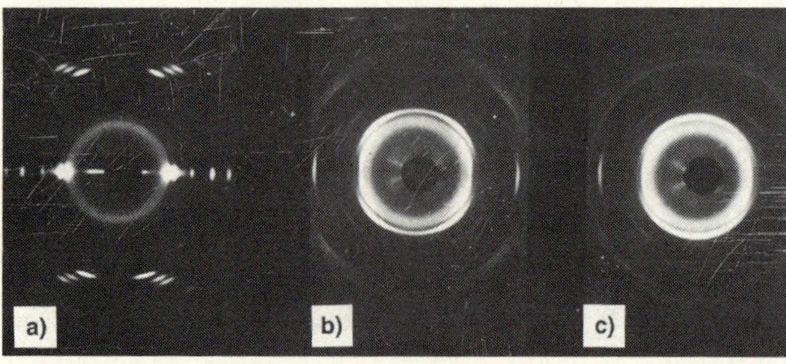


Fig. 15. X-ray diffraction patterns of polyethylene films crystallised under stress at three different stress levels (decreasing from left to right)

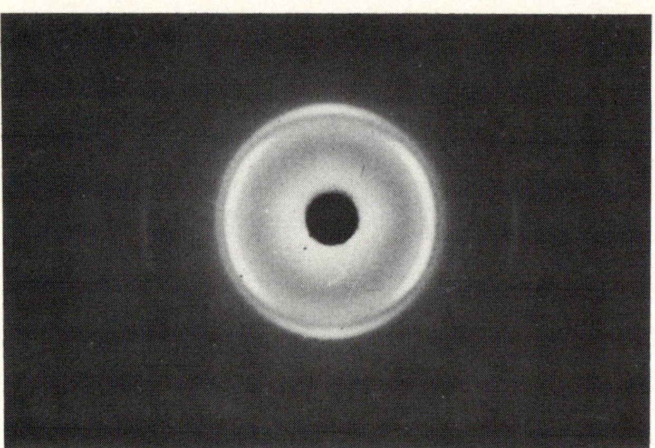

Fig. 16. X-ray diffraction pattern of a hauled off polyethylene film showing the "row structure" [22]

blown polyethylene films, is distinctly different from that of either a usual drawn fibre, or from that of a random film, yet it does not represent an intermediate state between the two, neither in terms of alignment of crystals nor in terms of molecular extension preceding crystallisation; it is due to a particular superposition of fully oriented fibres arising from oriented, stretched out molecules and of an essentially disoriented overgrowth, with a directionality which is determined merely by its epitaxial relation to the fibre but otherwise arising from initially random chains. In cases where the orienting influence is more complex, e.g. having a biaxial character such as in the blow moulding fabrication of films, the same consideration will still hold, except that here, as shown by Choi, Spruiell and White [24], the resulting multiaxial orientation is now in terms of the "rows" (Fig. 17) and not directly in terms of molecules and crystals, as without recourse to morphological background one may be tempted to envisage. Thus even in such a complex case the "oriented-unoriented" crystal texture or in molecular terms "extended and random" chain duality, our main theme here, still pertains.

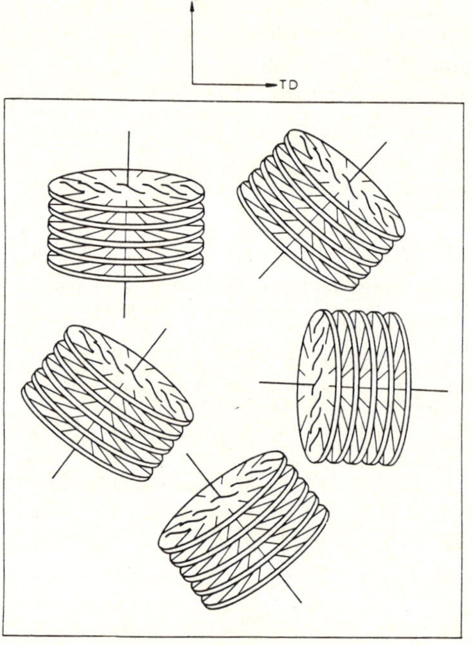


Fig. 17. Morphological model for a tubular blown film with equal biaxial orientation [24]

But even when confining ourselves to purely uniaxial situations, orientations which are intermediate between those in Fig. 15a and Fig. 16 can arise (Figs. 15b, 15c and sketched in Fig. 14e). In the beginning such orientations were held against the original "row structure" as intermediate orientations of crystals between *c* axis and *a* axis orientation (the latter by incorrect reading of diffraction patterns such as Fig. 16). However, it has become apparent [3, 23] that such orientations are intermediate only in the morphological sense and not in the sense of initial orientation of molecules and of the resulting crystals. Namely, the exclusive fibre-lamella duality, together with the preceding extended and random chain duality, our principal theme here, remains operative, merely the disposition of the lamellae will be intermediate between those in Figs. 14a and 14c giving rise to the "intermediate" diffraction patterns such as Figs. 15b and 15c and sketch 14e.

In morphological terms the above "intermediate" situations are best rationalised along the lines of Nagasawa et al. [25], as represented by Fig. 18 from their work. Here, the lamellar overgrowth starts with an epitaxial *c* axis alignment along the *c* axis oriented core fibre. As the lamellae grow outwards they gradually twist (or in general randomise) around *b* (in case of polyethylene). Full randomisation (around *b*) in terms of the resulting orientation is only reached after a number of turns of the lamellar crystal. As worked out quantitatively in ref. [25], by intergrating the orientation over cylinders of increasing diameter, the intermediate cases between pure *c* axis orientation (core fibre plus *c* axis aligned lamellar overgrowth close to the core) and the final "row orientation" (such as in Figs. 16 and 14b, corresponding to full randomisation around *b*, i.e. to the fully developed two dimensional spherulites) include the observed intermediate orientations such as Figs. 15b, 15c and Fig. 14e (including stages of predominant *a* axis orientation). Here the variable determining the final overall orientation is the cylinder diameter, which is determined by the distance between the fibres, in turn determined by the number of fibres themselves. I. E. many closely spaced fibres yield overall *c* axis orientation (constituted by fibres plus overgrowths), few, widely spaced fibres yield the "row" structure, essentially the orientation of the lamellae forming two-dimensional spherulites with spherulite planes perpendicular to the flow direction, and intermediate number of fibres allowing only partial randomisation of lamellar overgrowth give rise to orientations in between. All the above is again in full accord with our theme here, namely that all observed orientations are accountable in terms of a fibre-platelet duality as arising from the two extremes of extended and random molecules without intermediate stages in terms of chain extension and resulting crystal orientation, such as may be constructed from a picture of affine deformation of molecules and resulting gradual orientation of crystals. For a given material the orienting influence, namely the strength of the flow, only affects the fibre-lamella ratio, stronger flow creating more fibres with the consequences just laid out. For

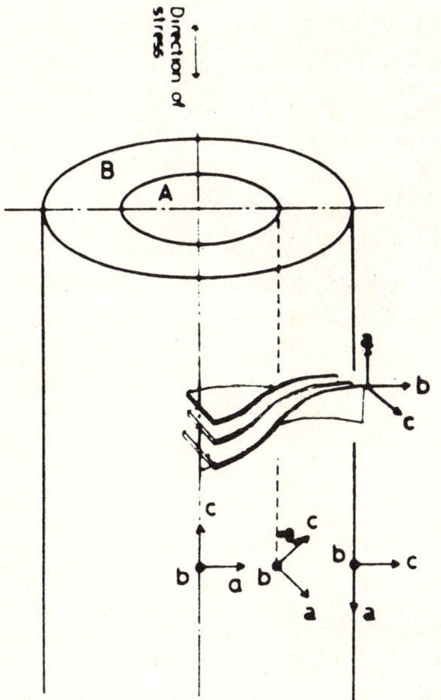


Fig. 18. Scheme due to Nagasawa et al. [25] to account for orientations intermediate between *c* axis and "row" orientation (such as in Fig. 15), including also predominant *a* axis orientation, in terms of increasing "row" diameter

a given flow strength the determining factor is the molecular weight: as will be referred to below higher molecular weight yields more fibres.

It may be worth pointing out that orientation parameters as derived say from birefringence without reference to the morphological picture are of little use in such situations, and can in fact be outright misleading. Similarly, the informative value of orientation functions derived from x-ray diffraction patterns may only be limited without the morphological background, and subsequent reduction of such functions to a single parameter, as often done, is to remove the clue to the true situation. However, orientation functions and parameters can again be informative when applied to the pertinent morphological frameworks as done by Nagasawa et al. [25].

### *A special case of fibre-platelet structures: interlocking shish-kebabs*

Finally, within this topic of fibre-platelet structures in melt crystallised material a special situation deserves separate mention. This is in the category of all *c* axis orientation, i.e. at the Fig. 14c end of the spectrum of columnar structures, with lamellae all parallel. Here, by appropriate control of the flow rate in a capillary rheometer and the cooling rate of the emerging extrudate, the lamellae can be obtained in a tapered form where the wedge-shaped lamellae from adjacent columns interlock, with actual examples shown by Figs. 19 and 20 and by the model Fig. 21 [31]. The origin of the tapering lies in the well known relation between lamellar thickness, i.e. fold length ($l$), and crystallisation temperature ($T_c$), namely $l \propto Q/\Delta T$ where $\Delta T = T_m^0 - T_c$ with $T_m^0$ being the equilibrium melting point and $Q$ containing the material parameters. It follows that when the temperature decreases in the course of crystal growth the lamellar thickness will decrease continuously according to the above relation leading to tapered overgrowths. Interlocking is achieved by suitable control of this taper and of the fibril separation. The latter is governed by the combined effects of strain rate (flow strength) and molecular weight according to the foregoings, the former by the conditions of cooling of the extrudate. The model in Fig. 21 was computer generated from the actual experimental input parameters and from the appropriate relations determining layer thickness and growth rates as a function of $\Delta T$.

Fig. 20. Enlargement of zip-fastener structure [31]

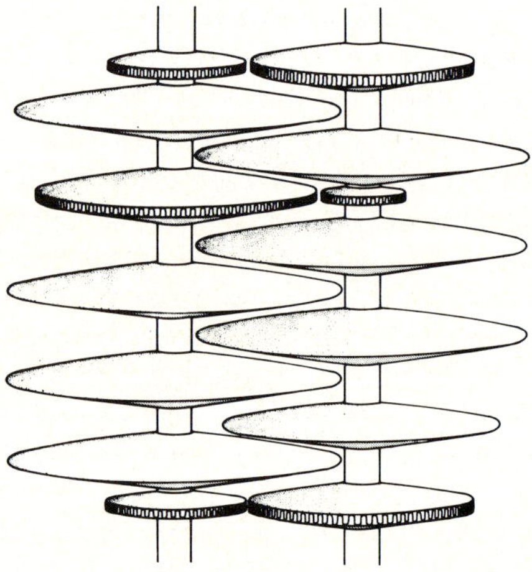

Fig. 21. Computer generated interlock model [26]

Fig. 19. Electron micrographs of interlocking shish-kebabs/zip-fastener structure (overall view), heavily stained to show up the larger scale columns (based on ref. [31])

The interlocking shish-kebab structures, having the appearance of a zip-fastener, in Figs. 20—21 are of interest from several points of view. They represent a special manifestation of the fibre-platelet duality theme of this paper: they show the potential of harnessing the morphology for specific purposes, and chiefly, the extrudates themselves, thus prepared, have special advantageous properties. The latter comprise high modulus (100 GPa by batch and up to 20 GPa by continuous processing) with low level of fibrillation and thermal shrinkage compared with high modulus material obtained along other routes.

## Analogy between fibre-platelet structures obtained from the melt and from solution

As will have become apparent the theme of fibre-platelet duality as the controlling factor of the orientation in flow crystallised product is being upheld throughout. By the foregoings the extended chain fibre-folded chain lamellar duality should be a reflexion of the coexistence of the fully stretched out chains and unextended chains in the appropriate flow system. The latter in turn is attributable to the coil → stretch transition in elongational flow, in the usual situation of a broad molecular weight distribution, where for a given strain rate the chains at the long end of the distribution become fully extended with the rest remaining unextended with no intermediate stage in between.

The above stated situation has been fully substantiated by the appropriate works on solutions: the basic molecular mechanisms of the coil → stretch transition and the ensuing shish-kebab formation in cases where chain stretching is followed by crystallisation. While there can be no corresponding conclusive studies on the melt, for reasons already stated, the observed effects, as described in the preceding section, strongly suggest that same argumentation can be carried over also to the melt.

To take them in turn, the effects all depend on elongational flow for their existence, from the original capillary blockage observations by van der Vegt to our own studies leading to the zip-fastener morphologies. In the case of the capillary rheometer the elongational character of the flow results from its convergence towards the orifice of the capillary die. In the case of commercial sheets containing the "row structure" and its variants the elongational flow is the consequence of the extension of the molten film in the course of processing.

The resulting crystal morphologies, as apparent throughout, display the same fibre-platelet duality which characterises the solution grown shish-kebabs. Further, the ratio of the two components of the dual structure is controlled by the same two variables as in case of solutions, namely strain rate and molecular weight, and this in the same sense, i.e. higher strain rate and higher molecular weight promote the formation of the fibre component. Admittedly, criticality in these two variables, such a salient feature of the solution studies, is not clearly apparent from the above quoted melt crystallisation effects, yet this would hardly be expected in view of the unavailability of closely monodisperse material for this kind of melt study. Even so, there is a striking sensitivity to variations in the high tail end of the molecular weight distribution. Thus, blending a minute amount ($\sim 1\%$) of ultra high $M$ material to a normal moulding grade produces conspicuous changes in the melt flow properties and consequent morphology: it can create sufficient number of fibres to produce the zip-fastener morphology, where otherwise such would not arise [26]. Thus it appears that in the elongational flow mode, such as through the orifice of a capillary extruder, the melt "is aware" of the fact that the small amount of added molecules are distinct from the rest, i.e. it "senses" their significantly longer length.

Nevertheless, the above mentioned parallel must have its limits, which is set by the strain which a given chain can experience in the kind of flow fields which exist in the capillary rheometer of the pertinent experiments. To recall, in the solution experiments the amount of strain sufficient for chain extension was assured by the presence of a stagnation point deliberately created by appropriate flow field design such as in the opposed jet geometry. Indeed, the effects arising between such opposed jets are identical both for solutions and melts (compare Figs. 10 and 11). Nevertheless, the single jet geometry can produce identical fibre formation (Fig. 13) from which similar happenings can be inferred for the analogous geometry at the entry orifice of the capillary rheometer. How the strain criterion can be satisfied in such a case, however, remains an open question. Calculations based on Newtonian flow cannot be relied on in a system which is so far from being Newtonian, and complex local deviations from idealised flow patterns can well be expected, expecially at the capillary entrance. While the issue cannot be resolved without experimental mapping of the actual flow fields in the pertinent experiments, the situation, as it stands, leaves the parallel between solution and melts at least open.

There is, however, a situation where the parallel between solution and melt must certainly cease. This is when stress induced crystallisation is statically induced, where nevertheless the morphological consequences seen are indistinguishable from those induced by genuine flow. It will be recalled that the verification of the "row structure" hypothesis, originally from observations on solidification during flow, was first substantiated by static experiments on lightly cross-linked polyethylene [23, 3]. In this case crystallisation was induced by stretching and, amongst others, followed in situ by x-ray diffraction, yielding patterns such as Figs. 14b, d, e, 15 and 16. Further, analogous electron microscope examinations on similarly stress crystallised, but lightly cross-linked thin films made the corresponding morphologies (such as corresponding to Figs. 14a, b and 18) directly visible. The same was accomplished by the extensive studies on stress crystallised elastomers of which Fig. 22 is an example [28, 29]. Clearly in all these static experiments the strains are far too low (at most a few hundreds of percent) to permit the extension of the entire chain. Neither is the extension of an individual chain expected under the circumstances when the system is indeed a mechanically connected network, be it, through chemical cross-links or through entanglements sufficiently permanent to maintain stress. Clearly, now we passed over from a situation such as Fig. 7 where $\dot{\varepsilon}_n > \dot{\varepsilon}_c$, to that in Fig. 8, where $\dot{\varepsilon}_n$ can be smaller than $\dot{\varepsilon}_c$. In the latter case the system will only be able to extend as a network, without a chance for the chains to extend individually as they may within the $\dot{\varepsilon}_n - \dot{\varepsilon}_c$ window which exists as long as $\dot{\varepsilon}_n > \dot{\varepsilon}_c$. We regard the issue of the alternatives, as expressed by Figs. 7 and 8 as possibly the most salient open ended question at present; specifically whether for $c = 1$ (single component melt) the situation depicted in Fig. 7 or Fig. 8 pertains, in other words, whether or not there exists a strain rate window where chains may become extended individually in the course of flow and, if so, under what conditions, or alternatively wether network deformations will necessarily always prevail. The present work is making a strong case for the situation in Fig. 7, but is also giving examples of statically stressed systems and lightly cross-linked rubbers, where the situation as in Fig. 8 must pertain, with morphological consequences which, at first sight, do not seem to distinguish between the two.

The last statement gives cause to some reflexion. By a priori considerations chain stretching by elongational flow in a flowing system and in a network are expected to be basically different. In the former it is the longest molecules which extend preferentially while in the latter it is the shortest network chains. If the latter are to act as crystal nuclei (a possibility raised in ref. [30]) they would be expected to generate crystals randomly distributed throughout the sample. This, however, is not observed: crystal nucleation always takes place along lines parallel to the stretch direction consistent with the formation of "row" or "shish-kebab" type structures, just as in crystals generated by elongational flow. Even the existence of a distinct central core fibre is sometimes just about visible as in Fig. 22 a distinct central line is conspicuously pre-

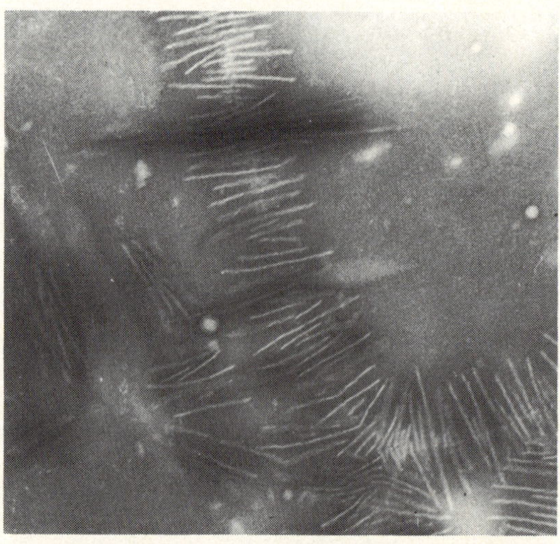

Fig. 22. Electron micrograph of a stress crystallised natural rubber showing the distinct central core of the fibre-platelet morphology (by courtesy of Prof. P. J. Phillips)

sent in shish-kebab type structures formed in strained isotactic polystyrene [32], which while not cross-linked has been strained, but not to such an extent as to allow appreciable chain extension by the elongational flow mechanism. Here, some kind of, presently unknown mechanism must be at play ensuring the linear progression of crystallisation along the initiating crystallisation direction. Beyond drawing attention to the parallel situation in as yet another work on isotactic polystyrene [33] we do not pursue this issue further. However, we cannot refrain from making a further point of potentially salient consequences which so far has not received due attention.

When starting from the point of view of stretching a permanent network, then by the whole extensive body of rubber elasticity the deformation is expected to be affine on the scale of our concern here (all the sophisticated arguments on how far the affine deformation criterion pertains relate to molecular dimensions such as junction distances and network chain lengths, but not to dimensions of several microns of our concern here, a dimensional range where the obeyance of the affine deformation criterion has never been questioned). However, we see from Fig. 22 and from numerous similar other micrographs in the literature (see review ref. [3]) that the rows of crystals are many microns apart even in polybutadiene and polyisoprene, the model materials for rubber elasticity. Thus, we not only have to face the issue as to why stress induced network crystallisation propagates along lines[3]) as opposed to having crystallisation centres statistically distributed throughout the sample, but also why these lines are so widely apart, this separation being on a microscopic rather than a molecular scale. Without entering detailed speculation the evidence as it stands indicates a highly inhomogeneous stress distribution throughout the film samples. Specifically, that the stress and the ensuing strain, such as generate crystallisation, is concentrated along lines microns apart giving rise (by whatever mechanism) to the central nucleating fibres, leaving the bulk of the material in between essentially unstressed, allowing it to crystallise in its usual, presumably chain folded, lamellar form. As far as the fibre-platelet duality is an indicator of the chain orientation in the preexisting amorphous state, the profound potential implications for rubber elasticity should be obvious: namely, that the stress distribution is not uniform, contrary to what has always been taken for granted, but is concentrated along the lines which are then made visible by subsequent crystallisation along these lines, or in other words, of which crystallisation serves as marker. What consequences this may have for the existing picture of a stressed rubber, with all it entails (stress-strain behaviour etc.) is not for us to say here.

## Rheological effects in the melt

### *Basic observations*

So far we have been concerned with the consequences of flow induced chain extension for the crystal morphology of the solidified sample, and through it for the type and strength of orientation in the final product. In this present and final section we shall describe some newly recognised highly unusual, and for applications potentially favourable effects in the behaviour of melts which we attribute to chain extension and chain extension induced phase transitions.

The set up where the effects are observed is again a capillary rheometer, operated at a constant piston velocity ($v$) and measuring extrusion pressure ($p$) as a function of temperature ($T$). Here $v$ relates to the shear rate at the capillary wall, most readily expressed as an "apparent wall shear rate" $\dot{\gamma}_A$, where

$$\dot{\gamma}_A = \frac{4Q}{\pi R^3} \qquad (2)$$

with $Q$ being material throughput per unit time, which is proportional to $v$. $R$ is the capillary radius and the ratio $p/\dot{\gamma}_A$ relates to the apparent flow resistance, hence apparent melt viscosity[4]).

The key effect, the basis of all what follows is shown in Fig. 23. Here, just as in the first experiments leading to the discovery of the effect in question [34—37], comparatively high $M$ material ($M = 10^5$—$10^6$) was used such as may display several kinds of extrudate distortion (Fig. 24; bottom) and unsteady flow at the usual processing temperatures (i.e. above 160 °C). However, rather surprisingly, at lower extrusion temperatures, between the temperature where solidification of the flowing polymer leads to blockage and roughly 152 °C, the flow becomes steady. In addition $p$ displays a pronounced minimum within a narrow temperature interval of 150—152 °C (Fig. 23) with the extrudate, which emerges, becoming smooth and uniform (Fig. 24; top). The effect is reversible with temperature, it takes place both on heating and on cooling (for qualifications pertaining to the cooling experiments see ref. [35]). Thus we have the rather surprising effect that material which would conventionally be unprocessable at such low extrusion temperatures, becomes potentially processable at temperatures below those used in conventional technological practice, with a narrow temperature window of minimum flow resistance ("extrusion window").

---

[3]) It is not self-evident that stress-induced crystallisation should propagate in a linear fashion unless the stress and the correspondingly strained molecules are localised along the lines in question already before crystallisation. Namely, crystallisation should lead to stress relaxation along the chain direction, which therefore will reduce the stress and not create more newly arising stress along the direction of crystal growth.

[4]) Following rheological practice we use here the symbol $\dot{\gamma}_A$ to denote apparent wall shear rate as rheologically defined, while retaining the symbol $\dot{\varepsilon}$ as the elongational strain rate experienced by the individual chain molecule. As we shall be concerned with anomalous flow conditions at which rheological correction schemes such as the Rabinowitsch-Weissenberg correction lose their validity, apparent $\dot{\gamma}_A$ values will not be corrected for to obtain real shear rates ($\dot{\gamma}$). Qualitatively, it follows that an increase in $v$ will have the result of increasing $\dot{\gamma}_A$, $\dot{\gamma}$ and, hence, $\dot{\varepsilon}$.

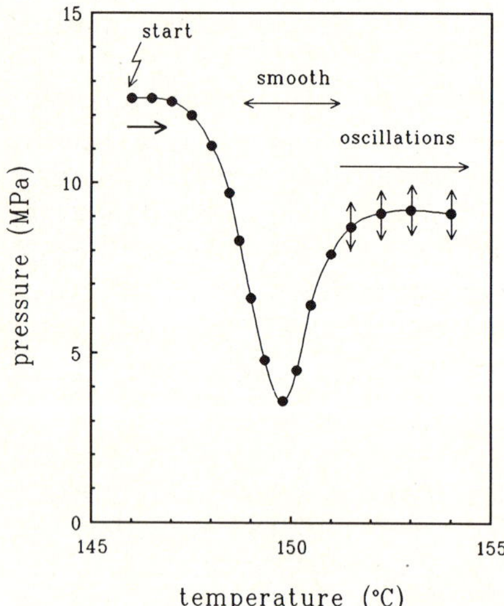


Fig. 23. Pressure vs. temperature trace showing the extrusion window. ($M$ = 440 000, $\dot{\gamma}_A$ = 1.89 $s^{-1}$) [36] (following ref. [34])

The above "window" effects were found to set in discontinuously above a certain shear rate ($\dot{\gamma}_{A,c}$) for a given molecular weight ($M$). This is shown in Fig. 25. Here for the lower $\dot{\gamma}_A$'s $p$ is a slowly decreasing function of $T$ without any sign of a minimum, as is expected from a steady decrease of melt viscosity with temperature ($\dot{\gamma}_A$ and the associated extrusion output here are very low so that flow instabilities, such as would characterise the extrusion at more practical speeds do not yet appear). However, the onset of a minimum in $p$ at 150—152 °C appears sharply on a small increment of $\dot{\gamma}_A$ beyond a certain $\dot{\gamma}_A$. Increasing $\dot{\gamma}_A$ further does not affect the location of the minimum along the temperature axis, but it makes the window more pronounced. We see therefore that the onset of the flow effect is critical in $\dot{\gamma}_A$.

Analogous criticality also applies to the molecular weights. We found that $\dot{\gamma}_{A,c}$ was strongly dependent on $M$: lower $M$ requiring a much enhanced $\dot{\gamma}_A$ for the minimum to set in, and vice versa. In spite of the fact that we did not possess sharp fractions for this purpose, only materials with comparatively broad $M$ distributions ($M_w/M_n \approx 5$), we could establish an apparently very well defined relation between $M_w$ ($\equiv M$) and the critical wall shear rate $\dot{\gamma}_{A,c}$ for the creation of the minimum in the $p$ vs. $T$ curve [36]. Accordingly, we found that:

$$\dot{\gamma}_{A,c} \sim M^{-4.0\pm0.1} \quad (3)$$

as derived from the double logarithmic $\dot{\gamma}_A$ vs. $M$ plot (Fig. 26).

Even the exact meaning of the —4 functional dependence apart, the message of Figs. 25 and 26 is clear: The appearance of the effect (the minimum in the $p$ vs. $T$ curve) is critical in both $\dot{\gamma}_A$ and $M$, the double criticality which is the hall mark of an elongational flow induced coil → stretch transition.

The same criticality is strikingly confirmed by extrusion experiments carried out at constant $p$. Here,

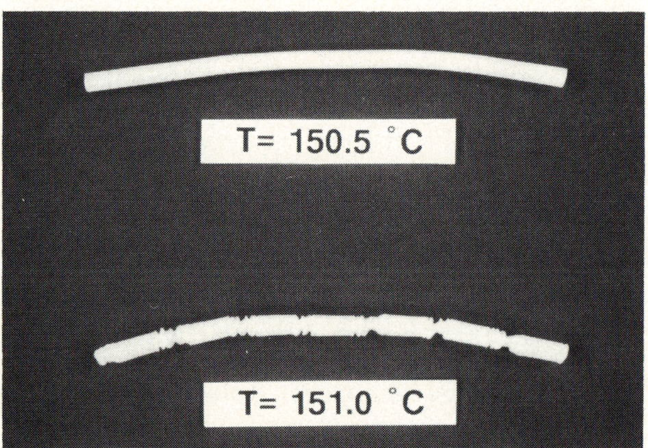


Fig. 24. Photograph of extrudate ($M$ = 300 000, $\dot{\gamma}_A$ = 3.78 $s^{-1}$). Top: $T$ = 150.5 °C (smooth). Bottom: $T$ = 151.0 °C (rough) [36]

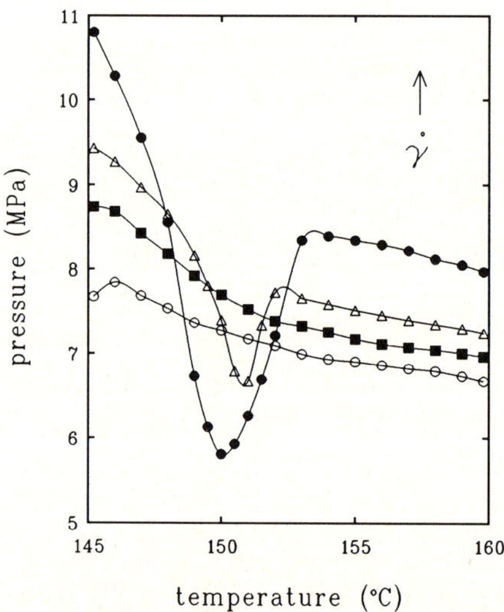


Fig. 25. Pressure vs. temperature traces showing the onset of the extrusion window. From bottom to top: $\dot{\gamma}_A$ = 2.27, 2.65, 3.02, 3.78 $s^{-1}$. $M$ = 300 000 [36] (following refs. [34] and [35])

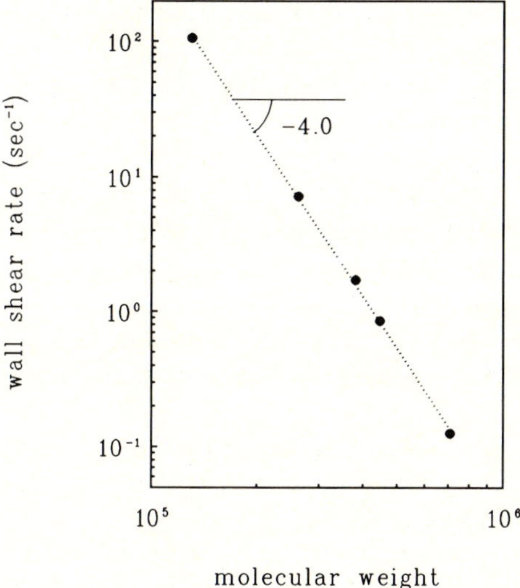


Fig. 26. Critical apparent wall shear rate ($\dot{\gamma}_{A,c}$) at the onset of the extrusion window vs. $M$ [36]

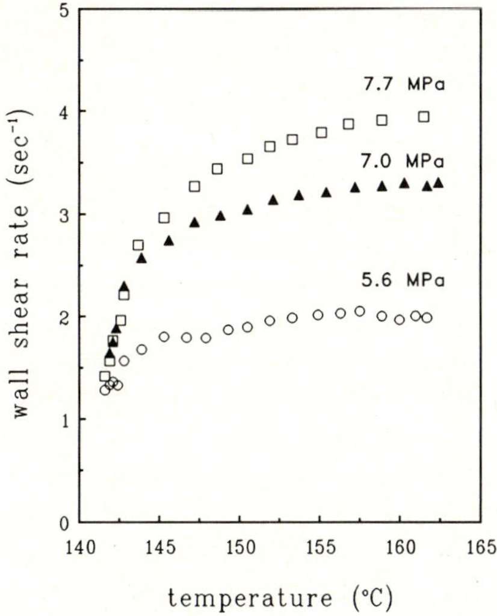


Fig. 27. $\dot{\gamma}_A$ as a function of $T$ for different values of constant $p$ under normal extrusion conditions. $M$ = 300 000 ($p$-levels are indicated) [36]

(using a Göttfert Rheograph 2002 instrument), the pressure $p$ was kept fixed while lowering the temperature with the piston displacement rate ($v$) being measured. The experimental results are displayed by the $\dot{\gamma}_A$ vs. $T$ curves of Figs. 27 and 28 obtained at increasing values of constant $p$. In Fig. 27 $\dot{\gamma}_A$ is an increasing function of $T$, as to be expected from the increasing fluidity (or decreasing viscosity) of the melt with $T$. As can be observed, $\dot{\gamma}_A$ (or $v$) values are higher for higher values of $p$ (the curves are lying above each other, in the sequence of increasing $p$) in line with expectation of higher shear rates in response to higher applied stresses. So far it is all normal rheological behaviour. However, as shown in Fig. 28 the situation changes dramatically on further increase of $p$: at a specific $p$ the $\dot{\gamma}_A$ vs. $T$ curve shows a sharp maximum at 150—151 °C. On further increase in $p$ the position of this maximum in $\dot{\gamma}_A$ remains unaltered but increases in size. (For $p$ levels above those listed in Fig. 28 the pressure could no longer be maintained at its fixed value by the apparatus, because the piston was travelling too fast, with concomitantly fast squirting out of the extrudate, to be followed instrumentally).

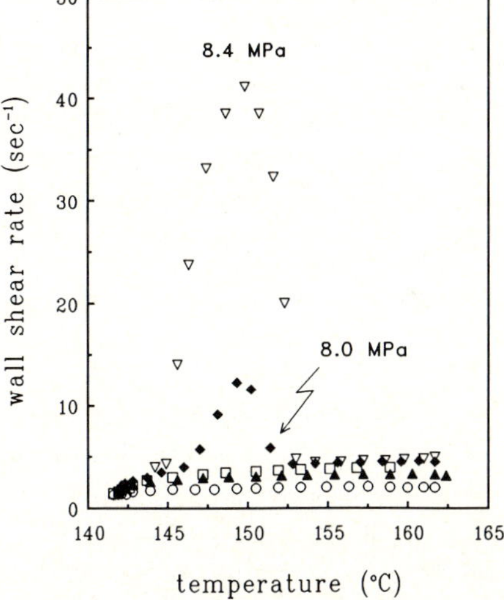


Fig. 28. $\dot{\gamma}_A$ as a function of $T$ for different values of $p$, both under normal extrusion conditions (bottom curves, copied from Fig. 27) and under conditions uncovering the extrusion window (top curves). ($p$-levels are indicated) [36]

As will be apparent the above experiment as represented by Figs. 27 and 28 is the precise complement of that in Fig. 25, both registering the same effect. The first in the form of $p$ as a function of $T$ at constant $v$ (hence $\dot{\gamma}_A$), the latter in the form of $v$ (hence $\dot{\gamma}_A$) as a function of $T$ at constant $p$. In the first case the effect is manifest in the form of a minimum and in the latter as a maximum in the respective curves vs. $T$. We interpret both effects as manifestations of criticality in $\dot{\gamma}_A$, in Fig. 25 directly and in Fig. 28 indirectly via $p$, where, by our interpretation $\dot{\gamma}_A$ increases with $p$ until its critical value

$\dot{\gamma}_{A,c}$. Ongoing work indicates that the critically in $M$ is also mirrored by both types of experiment with constant $v$ and also with constant $p$.

The next question is the reason for the effect appearing only within a narrowly defined temperature interval, and why precisely in the window of 150—152°C. Clearly, such a sudden change in rheological behaviour cannot arise from rheological considerations alone. Such an effect must arise from a change in structure, hence ultimately its origin must be in thermodynamics, i.e. it must lie in a phase change. Chain extension would be precondition for such a phase change with the latter then setting in discontinuously at the temperature appropriate for the phase transition. Regards the type of phase transition it is not likely to be crystallisation. Crystallisation is indeed induced by chain extension, in fact this was the subject of the preceding chapters in this paper dealing with the morphological consequences of such a crystallisation. However, crystallisation is impeding and not promoting the flow as manifest the complete blockage in the original van der Vegt [20] and Porter [21] experiments. Indeed, the upswing of the $p$ vs. $T$ curves in Fig. 23 and the downturn of the $\dot{\gamma}_A$ vs. $T$ curves in Figs. 27 and 28 towards the low temperature end are the manifestations of such crystallisation which at still lower temperatures leads to complete blockage of flow. In fact all the flow-induced crystalline structures described earlier arise, in association with the upturn in the $p$ vs. $T$ curve towards lower temperature, where crystallisation does occur, without yet preventing continuous passage of material through the capillary extruder. It follows that the new phase responsible for the dip at ~151°C in the $p$ vs. $T$ curve must be of a different kind. We suggest that this is a phase intermediate between the melt and the crystal. In PE such a phase could be the hexagonal phase known to exist from other sources.

In the stationary state, and under ambient conditons, the hexagonal ($h$) phase is metastable in relation to the orthorhombic ($o$) crystal, the stable form of solidified PE. Nevertheless, as is familiar, the $h$ phase can become the stable one under high hydrostatic pressure [38] through which it is known to be highly mobile with liquid crystal type characteristics, which is the property we would require for explaining our rheological effect. Also, the $h$ phase can be produced by heating constrained stretched fibres when the $o$ form instead of melting transforms into the $h$ form [39, 40]. Here, the reason for activating the otherwise metastable $h$ phase is the lowering of the entropy (and consequent raising of the free energy) of the molten state by preserving the chains stretched out even in the melt, which as can be argued from a priori considerations [41], will "uncover" a metastable phase intermediate between the melt and the crystal of ultimate stability. As will be noted this is already close to the present situation in flowing melts where we rely on the flow to stretch out the chains and keep them, even if merely transiently, in the stretched state, when given the appropriate temperature, they could transform into the $h$ phase. It is probably no coincidence that the $o \rightarrow h$ transition in the constrained fibre experiment is 150—151°C [39, 40], i.e. precisely the temperature of the pressure minimum in the present rheological experiment. All this is in full support of our proposed explanation of the rheological window being due to an elongational flow induced coil → stretch transition which in turn generates a highly mobile phase at a specific temperature.

*Implications*

At this point we may take stock of this situation and of its implications. The combined criticality of the extrusion window with respect to $\dot{\gamma}$ and $M$ forms a parallel to our experience with the stretching of individual chains in our controlled elongational flow experiments in solutions, in full accord with the critical expectations of a coil → stretch transition. Thus we have the clear message on the existence of such a coil → stretch transition also in the melt, even for the highest molecular weights (up to $10^6$) here considered, difficulties this may raise from the traditional view point of melt behaviour notwithstanding. It follows, that, as far as the flow field is concerned, this can "see" the individual molecule as such even in the melt, otherwise it could not discriminate between chains of different lengths in such a clearly delineated way.

All the above supports our main theme in this paper, the "on-off" effect of chain stretching and orientability both in solution and in the melt with all its manifold consequences for structure and orientation within the resulting solids, and as we now see, even for certain flow effects (the rheological "window") in the melt. Significantly, the invariance of the window temperature with shear rate (hence strain rate) is also explicable in terms of our scheme, in fact it is strongly supportive of an on-off effect with no intermediate stage in between. That is, increasing the strength of the flow does not increase the orientation or extension of the chains themselves in any substantial manner, but merely creates more stretched out material in the sense outlined in the earlier sections. Accordingly, the $p$ minimum will be more pronounced, but will not be displaced along the $T$ axis as indeed observed. Thus we are not increasing the chain extension and the resulting phase transition temperature in any continuous manner through the applied field: we only have two states, the conformationally random and the conformationally fully stretched out chain where, in the combined presence of a range of molecular lengths and of an, in practice, unavoidably inhomogeneous flow field, the relative amounts of the two but not the nature of each is being affected by the strength of the field[5]).

The existence of a coil → stretch transition in the melt, so strongly suggested by the present experiments, would raise important questions regards the nature of melt flow itself. In solutions it is being envisaged as the consequence of the screening effect of the chain. Namely, in the non-free draining coil state the coil interior is being screened from the effect of flow in the surrounding solvent and the run-away effect of the coil → stretch transition on increasing strain rate (with no intermediate stage as a steady state, except as a transient) is the consequence of the coil interior becoming increasingly exposed to frictional contact with the solvent, as it starts opening up beyond a critical strain rate. As far as the same effect would be due to the same cause also in the melt one may look for the reason and the origin of analogous screening effects there as well. As a suggestion, this could be provided by a high degree of polydispersity in usual polymer melts. Accordingly, a molecule at the high tail end of the distribution could be regarded as a long molecule in a solution of its shorter species which would provide a screening effect analogous to that in a solvent. The indicated response to elongational flow, including the coil → stretch transition, would then follow. Possibly, the general experience that polydispersity is favourable to steady stretching flows while the trend towards monodispersity is inducive to flow instabilities (network response [43, 44]) is supportive of the above argument.

## Conclusion

This article provides a survey on flow induced chain extension, both in terms of fundamental principles and morphological consequences, such as to our knowledge have not been brought together under the same unifying umbrella previously. The overall conclusion which emerges is that a wide range of phenomena is accountable on the basis of a dual population of chain orientation, namely essentially random and practically fully extended chains within the same sample volume. The effects in question range from morphologies and x-ray registered orientations of melt solidified products, to recently observed singularities in melt flow rheology displaying, in our interpretation, the consequences of the two extremes of chain orientation as coupled with liquid → solid phase transitions appropriate to each. The above considerations, while internally self consistent, focus on further open

[5]) The above is clearly an approximation which we feel is justified for our present purpose. Obviously the random coil state will exist in different states of deformation, which here we consider collectively as "slight" and hence ignore with respect to the extended state. Also, there will be degrees within what we regard as fully extended chain category. These again we ignore, by considering all configurations close to full chain extension collectively as extended, as at that advanced stage of extension the variants will only have little effect on the resulting entropy. This is not to say, however, that differences in stretching of valence bonds will not have a significant effect on the free energy, hence on the phase transition temperature. We maintain however, that this stage of chain stretching will not be reached to any appreciable extent in the course of the preset flow experiments, where, as we envisage, the chains are being stretched out individually and not as part of a network. Namely, in such individually stretched out chains the stress concentration is exactly in the cenre: any overloading will break the chains there (precise central cleavage demonstrated conclusively in this laboratory [8]) instead of stressing the rest of the bonds along the chain in any substantial manner, which would be needed to shift the transition temperature. The latter in fact would be accomplished by applying the stress uniformly along the fully stretched out chain by embedding ultra oriented fibres in a resin as was done by Rastogi and Odell [42] who observed a raised $o \rightarrow h$ transition temperature up to 164°C. Even if no additional stress was applied from outside, the prevention of shrinkage, which should occur on an $o \rightarrow h$ transition (the $c$ spacing is smaller in the $h$ than in the $o$ phase), should have the same effect.

ended issues concerning similarities and differences between extensibility of free chains and networks and on apparently large scale inhomogeneities in stress distribution which may appear in the latter.

*Acknowledgement*

Financial support by DSM is gratefully acknowledged by one of us (J.W.H.K.).

**References**

1. Keller A (1979) In: Ciferri A, Ward IM (eds) Ultra-High Modulus Polymers. Appl Sci Publ London, p 321
2. Keller A (1977) J Polym Sci Polym Symp 58:395
3. Keller A, Mackley MR (1974) Pure Appl Chem 39:195
4. de Gennes P-G (1974) J Chem Phys 60:15
5. Peterlin A (1966) J Polym Sci B4:287
6. Pope DP, Keller A (1978) Coll Poym Sci 256:751
7. Odell JA, Narh KA, Keller A (1992) J Polym Sci: Part B: Polym Phys 30:335
8. Odell JA, Keller A, Muller AJ (1992) Coll Polym Sci 270:301
9. Keller A, Odell JA (1985) Coll Polym Sci 263:181
10. Mackley MR, Keller A (1975) Phil Trans Roy Soc London Ser A 278:29
11. Miles MJ, Keller A (1980) Polymer 21:1295
12. Odell JA, Keller A, Miles MJ (1985) Polymer 26:1219
13. Keller A, Muller AJ, Odell JA (1985) Progr Coll Polym Sci 75:179
14. Chow A, Keller A, Muller AJ, Odell JA (1988) Macrom 21:250
15. Pennings AJ, van der Mark JMAA, Booij HC (1970) Kolloid Z v Z Polymere 236:99
16. Narh KA, Keller A, unpublished results
17. Pennings AJ (1992) Lecture at EPS Conf St Petersburg
18. Mackley MR, Keller A (1973) Polymer 14:16
19. Mackley MR, Frank FC, Keller A (1975) J Mater Sci 10:1501
20. van der Vegt AK, Smit PPA (1967) Adv Polym Sci Monograph 26, Soc Chem Ind London 313
21. Southern JH, Porter RS (1970) J Macromol Sci B4:541
22. Keller A (1955) J Polym Sci 15:31
23. Keller A, Machin MJ (1967) J Macromol Sci B1:41
24. Choi KJ, Spruiell JE, White JL (1982) J Polym Sci Polym Phys Ed 20:27
25. Nagasawa T, Matsumura T, Hoshino S (1973) Appl Polym Symp 20:295
26. Bashir Z, Odell JA, Keller A (1984) J Mater Sci 19:617
27. Hill MJ, Keller A (1969) J Macromol Sci-Phys B3(1):153
28. Andrews EH (1966) J Polym Sci A-2, 4:663
29. Phillips P (1983) chapter 2 of Engineering Dielectrics, Vol IIA, Bartnikas R, Eichhorn RM (eds) ASTM STP 783
30. Jenkins H (1974) Ph D thesis, University of Bristol
31. Odell JA, Grubb DT, Keller A (1978) Polymer 19:617
32. Dlugosz J, Grubb DT, Keller A, Rhodes MB (1972) J Mater Sci 7:142
33. Petermann J, Miles M, Gleiter H (1979) J Polym Sci: Polym Phys Ed 17:55
34. Waddon AJ, Keller A (1990) J Polym Sci: Part B: Polym Phys 28:1063
35. Narh KA, Keller A (1991) Polymer 32:2513
36. Kolnaar JWH, Keller A, Polymer in the press
37. Waddon AJ, Keller A (1992) J Polym Sci: Part B: Polym Phys 30:923
38. Bassett DC, Block S, Piermarini GC (1974) J Appl Phys 45:4156
39. Clough SB (1970) J Macrom Sci B4:199
40. Pennings AJ, Zwijnenburg A (1979) J Polym Sci: Polym Phys Ed 17:1011
41. Keller A, Ungar G (1991) J Appl Polym Sci 42:1683
42. Rastogi S, Odell JA (1993) Polymer 34:1523
43. Vinogradov GV, Malkin AY, Yanovskii YG, Borisenkova EK, Yarlykov BY, Berezhnaya GV (1972) J Polym Sci A-2, 10:1061
44. McLeish TC, Ball RC (1986) J Polym Sci: Polym Phys Ed 24:1735
45. Hill MJ, Keller A (1981) Coll Polym Sci 259:335

Authors' address:

Prof. A. Keller
H. H. Wills Physics Laboratory
University of Bristol
Tyndall Avenue
Bristol BS8 1TL, United Kingdom

**Note added in proof**

Since this paper has been submitted our views on the rheological effects (last chapter) have changed in the light of further results. The effects described here have been further consolidated and so has the model of chain extension induced phased transformation underlying the singularity in flow behaviour. Nevertheless, contrary to the crystallisation and morphology related effects, such as blockage, fibre-platelet duality etc., which are located at the capillary entry, the site of the rheological effect is now identified as being along the capillary wall. This unexpected recognition opens up new unforeseen perspectives in the subject of melt flow and associated chain extension, but at the expense of some of the ties between the last section and the rest of the paper envisaged originally.

**Progress in Colloid & Polymer Science** Progr Colloid Polym Sci 92:103—110 (1993)

# New developments in the production of high modulus and high strength flexible polymers

I. M. Ward

Interdisciplinary Research Centre for Polymer Science and Technology, University of Leeds, Leeds, United Kingdom

*Abstract:* Recent research at Leeds University on the production on highly oriented flexible chain polymers is summarised. For melt spun polyethylene fibres, particular attention has been given to tensile strength, creep behaviour and surface treatments to improve adhesion to resins. Recent research on hybrid composites and 100% polyethylene fibres composites is summarised. A die drawing process has been developed for the production of oriented rod, sheet and tube, including biaxially oriented tube. This process is applicable to polyethylene, polypropylene, polyvinylchloride, polyester and other polymers.

*Key words:* High modulus, high strength flexible polymers — die drawing process — polyethylene

## Introduction

The last 20 years has seen the development of high modulus and high strength polymers by two principal processing routes: 1) spinning a mesophase in a lyotropic or thermotropic liquid crystalline polymer, and 2) imposition of very high degrees of plastic deformation in a flexible conventional crystalline polymer.

The first route of spinning a mesophase has been exemplified by the development of the lyotropic poly(p-phenylene terephthalamide) (Kevlar and Twaron fibres) and the thermotropic copolyester of hydroxybenzoic acid and 2—6 hydroxynaphthoic acid (Vectran fibres). These fibres derive their high stiffness and strength from their intrinsic chain stiffness and rigidity, and the ability to produce very high degrees of molecular orientation in the final fibre primarily due to high alignment of the mesophase during spinning. The second route to high modulus and high strength fibres has been exemplified by polyethylene fibres, either by melt spinning (Tenfor) or by gel spinning (Dyneema and Spectra) followed by aligning and unfolding the molecular chains by stretching to very high draw ratios.

Most of these developments have been concerned with the production of fibres, for practical reasons (either recovery of solvent or the requirement of very high deformation rates) in the case of the liquid crystalline systems. It has, however, been appreciated since the initiation of these developments that the solid phase deformation route should be applicable for the production of high modulus and high strength polymers in large section. Originally this was explored by ram extrusion and hydrostatic extrusion, and very spectacular results have been obtained, especially for polyethylene. These solid phase extrusion processes are batch rather than continuous processes and require comparatively low production rates. Recently a new continuous process operating at more realistic rates has been developed. This is the die-drawing process, which offers considerable new possibilities for biaxially oriented tube and sheet.

In this paper, discussion will be confined to the production of highly oriented flexible polymers by solid phase deformation processes.

## High modulus polyethylene fibres

One of the key results which underpins the development of high modulus flexible polymers is the simple guide-line that the Young's modulus depends to a good first approximation only on the draw ratio. The establishing of this guide-line for polyethylene by Andrews and Ward [1] in 1970, followed a considerable body of research by the present author and his colleagues in ICI during the period 1954—1965. The key ideas were 1) The Young's modulus of fibres (polyethylene terephthalate, nylon, polyethylene, polypropylene, etc.) depends primarily on molecular orientation, with crystallinity and morphology playing only minor roles). Mathematically this was formulated by the aggregate model [2]. 2) The molecular orientation can be related to the draw ratio by various deformation schemes, of which the affine rubber network model and the Kratky pseudo-affine rotating rod model are the best known [3].

The target of high modulus polyethylene fibres was identified because it was recognised that the polyethylene chain modulus was very high. This had been shown theoretically by Treloar [4], and measured from x-ray crystal strain results on fibres under stress by Sakurada and his co-workers [5]. Furthermore it was appreciated that the quest for high modulus should be translated into a quest for a very high draw ratio, i.e. to control the initial morphology of the polyethylene by optimising molecular weight and morphology to give a molecular network of exceptional extensibility. This was achieved by the research of Capaccio and Ward [6] (Fig. 1) so that draw ratios of 30 and room temperature Young's moduli of 70 GPa are now quite commonplace for oriented polyethylene fibres and tapes.

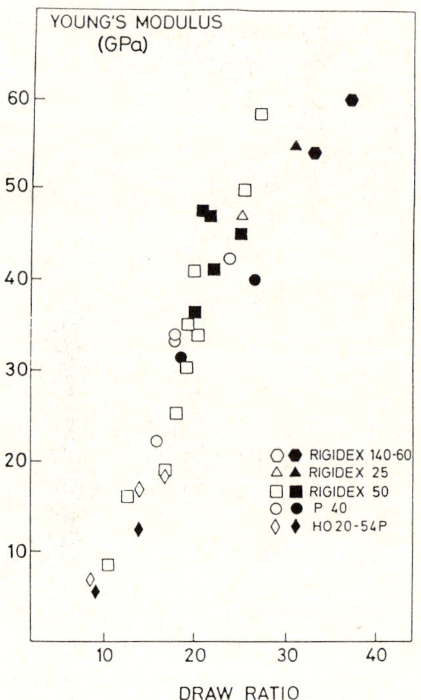


Fig. 1. Room temperature Young's modulus versus draw ratio for quenched (open symbols) and slow cooled (solid symbols) linear polyethylene samples drawn at 75°C. (Reproduced from J. Polym. Sci., Polym. Phys. Ed. (1976) 14, 1641 by permission of the publishers, John Wiley & Sons Ltd. (C).)

In later research high modulus and high strength polyethylene fibres were produced from ultra-high molecular weight polyethylene by Zwijnenburg and Pennings [7] and by Smith et al. [8]. In a development which has led to commercially produced fibres, Smith and Lemstra [9] recognised that a highly extensible network could also be produced by spinning a gel structure of high molecular weight polymer to reduce the number of physical entanglements. Because the strength of an oriented fibre is dependent on molecular weight (a well-known result from fibre technology) the gel spun fibres are substantially stronger than melt spun fibres of comparable draw ratio.

The factors which determine the strength of polyethylene fibres have been the subject of extensive investigation in recent research at Leeds University [10]. Most of the studies have been concentrated on melt spun fibres but comparison has been made with gel spun fibres. It is important in the first instance to attempt to determine the extent to which the fibre strength relates to macroscopic factors such as flaws and fibre diameter, or microscopic and structural factors such as molecular weight and orientation. For melt spun fibres it was possible to produce fibres of markedly different diameters at comparable draw ratio and modulus. Following standard procedures the Weibull moduli were determined for the fibres. A significant diameter effect was found at high draw ratios, and the Weibull moduli obtained from the distribution of fibre strengths agreed well with that obtained from the diameter effect on the basis of either surface flaws or volume flaws depending on the details of the processing route. For low draw ratios the effects of molecular weight and molecular weight distribution were explored in some detail. The results were surprisingly consistent with the proposal originally due to Flory [11], that strength $\sigma$ relates to the number average molecular weight $\bar{M}_n$, by a relationship of the form

$$\sigma = A - B/\bar{M}_n ,$$

where $A$ and $B$ are constants.

Extrapolation of data for a range of samples of different polydispersity showed that the tensile strength of monodispersed polymers could be described by

$$\sigma = KM^{0.25} ,$$

where $K$ has values 68 MPa and 78 MPa for draw ratios 15 and 20 respectively. Moreover, the tensile strengths of polydispersed polymers could be predicted by assuming that the tensile strength relates to the weighted average of the strengths of the different molecular weight components, i.e. a simple weight average summation. This weight average summation rule is also predicted by a fracture model proposed by Termonia et al. [12] which assumes that random bond breaking occurs as stress is transmitted through the network structure of chains of different length randomly distributed within the fibre. This model has been confirmed by their results for both gel spun and melt spun drawn fibres.

It is, however, necessary to point out that this model, which incorporates the proposal that strength depends on molecular weight according to the Flory scheme, assumes that the strength is limited by molecular factors. Measurements of the Weibull modulus on highly drawn fibres (either gel spun or melt spun) show values consistent with the conclusion that the strength is flaw limited. It must be concluded, therefore, that both intrinsic factors (molecular weight and orientation) and extrinsic factors (flaws) are important, so that the processing conditions can play a major role in the final strength. This is very well illustrated by Fig. 2, which shows how the tensile strength (tenacity) of melt spun fibres is dramatically affected by draw temperature at high draw ratios.

A major feature of polyethylene fibres is their viscoelastic nature and the possibility of permanent plastic deformation when subjected to stresses which are a comparatively small fraction ($\sim 0.1$) of their failure stress. The creep behaviour has been the subject of extensive investigations at Leeds University [13], and a methodology was developed which has proved useful for critical evaluation of performance. It was appreciated that the greatest physical understanding of creep processes can be obtained if the creep rate can be studied under quasi-equilibrium conditions, and following the work of Sherby and Dorn [14], if creep rate is examined as a function of creep strain. The resulting

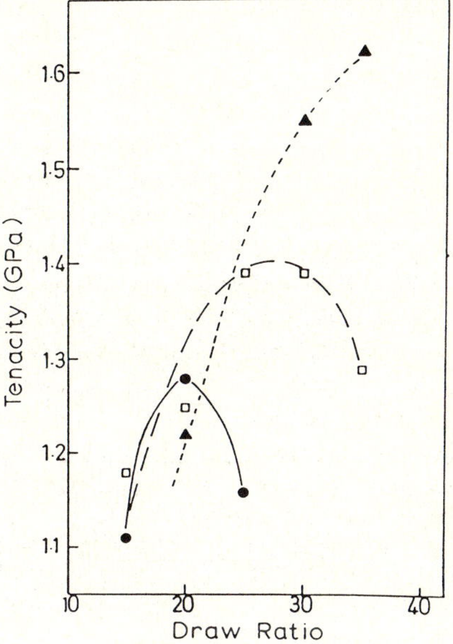


Fig. 2. The effect of draw ratio on —55°C tensile strength as a function of draw temperature on samples of Alathon 7030 monofilament. (●) $\lambda_{DT}$ = 100°C; (□) $\lambda_{DT}$ = 115°C; (▲) $\lambda_{DT}$ = 125°C. (Reproduced from J. Mater. Sci., (1986) *21*, 4199 by permission of the publishers, Chapman & Hall Ltd. (C).)

plots for polyethylene (Fig. 3) are quite dissimilar from those of Sherby and Dorn for polymethylmethacrylate in that after an initial period where very fast creep occurs the creep rate becomes constant with time (or strain) so that an equilibrium plateau creep rate is observed. For convenience, we have continued to construct Sherby-Dorn plots but it is important to appreciate that to obtain a physical understanding it is necessary to plot the plateau creep rates as a function of stress (Fig. 4). These plots then enable us to make a crucial evaluation of different materials for engineering purposes, because, as shown in Fig. 4, there is a region where the creep rate falls dramatically with decreasing stress thus defining a critical stress (as first pointed out by Wilding and Ward [13]) below which the creep rate is negligible. This critical stress can therefore be used to define a safety factor for engineering purposes, and this has been done very successfully (with the present author's guidance) for Tensar [15].

The key practical results from the creep studies show that the creep performance relates to the following factors

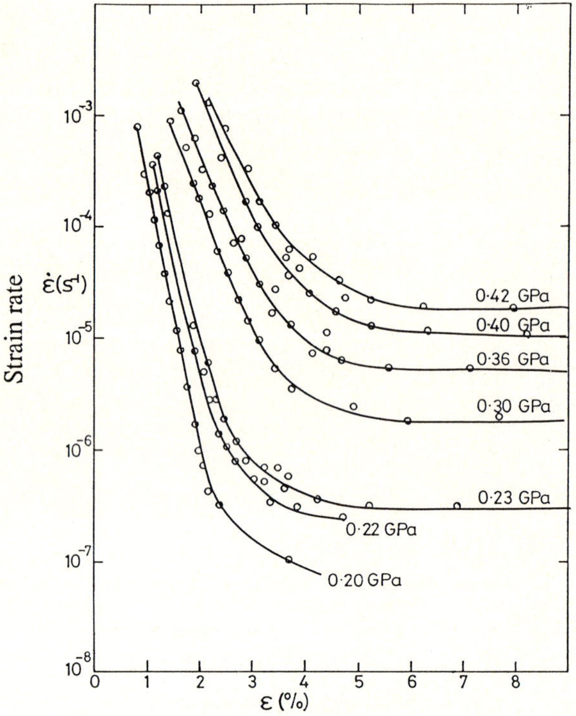


Fig. 3. Sherby-Dorn plots for a typical sample (HO20, $\lambda$ = 20) at indicated stress levels. (Reproduced from J. Polym Phys. Polym. Phys. Edn., 1984, *22*, 561 by permission of the publishers John Wiley & Sons. Inc. (C).)

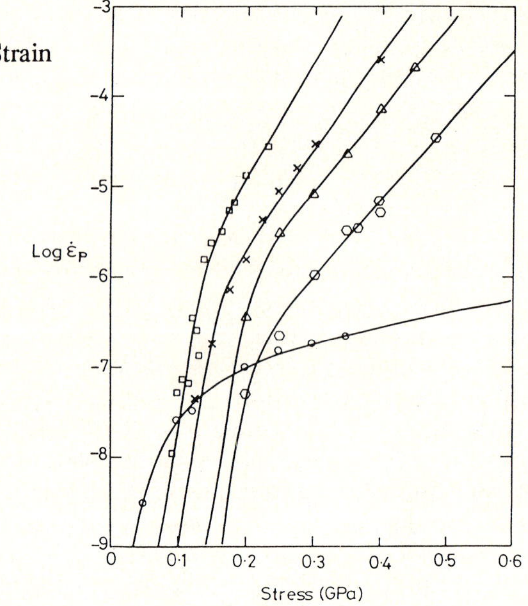


Fig. 4. Curves fitted to plateau creep data on the bais of the two-process model, assuming that process 1 has an activation volume of 456 Å$^3$. (□) Rigidex 50, $\lambda$ = 20, (X) 006-60, $\lambda$ = 20, (△) $\gamma$-irradiated Rigidex 50, $\lambda$ = 20; (○) HO20, $\lambda$ = 20; (●) Hostalen GUR, solution-spun fibre. (Reproduced from J. Polym. Phys. Polym. Phys. edn., 1984, *22*, 561, by permission of the publishers John Wiley & Sons Inc. (C).)

1) Draw ratio, creep being substantially reduced with increasing draw ratio.
2) Molecular weight, creep reducing with increasing molecular weight.
3) Branch content, creep reducing with increasing branch content.

It has been shown that creep rate/stress plots of the form obtained (Fig. 4) can be very well described by a two-process model with two thermally activated processes acting in parallel. It has been proposed that there is a high stress low activation volume process associated with crystalline c-slip and a low stress high activation volume process associated with the molecular network. This led to the invention of reduced creep fibres by irradiation cross-linking [16], either by x-rays or $\gamma$-rays. It was found that dramatic improvements in the effectiveness of irradiation cross-linking were obtained by irradiation in an atmosphere of acetylene, followed by appropriate annealing at high temperatures. Typical results are shown in Fig. 5.

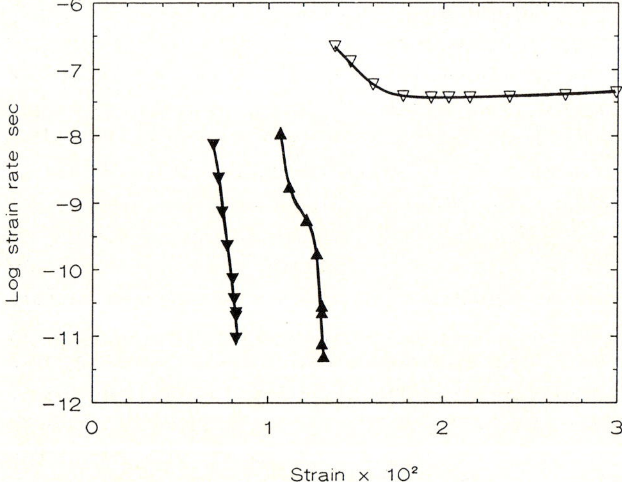


Fig. 5. Creep rate for melt spun high modulus polyethylene fibres. $\gamma$ irradiation in acetylene, annealed in acetylene, 0.17 GPa stress applied at 20°C: ▽ untreated, 0% gel; △ 3 MRad dose, annealed 1 hour at 110°C, 75% gel; ▼ 6 MRad dose, annealed 1 hour at 80°C, 82% gel. (Reproduced from Plastics and Composite Processing and Applications Rubber, ((1985) 5:157), by permission of the publishers Elsevier Science Publishers (C).)

Recent research suggests that the acetylene promotes the production of diene cross-links between the chains and reduces any tendency for chain-scission [17]. The essential idea is that the cross-linked network takes a substantial proportion of the applied stress, thus reducing the creep rate and also provides a mechanism for recovery.

The high chemical resistance of polyethylene fibres is a positive virtue for many applications, but it also implies that without appropriate surface treatment the bonding of the fibre to a resin matrix in a composite will be inadequate. There have therefore been systematic effects to devise surface treatments which provide improved fibre/resin adhesion. Fundamental studies were undertaken using large diameter (~260 μm) drawn monofilaments, monitoring the adhesion by a simple pull-out test where the load required to extract the fibre from a thin disc of cured resin (usually a standard epoxy resin) was measured. Following a wide survey of chemical treatments (e.g. chromic acid) and plasma treatments with a range of gases under different conditions, it was shown that a major improvement in adhesion could be obtained by plasma treatment in oxygen [18]. Three separate mechanisms have been identified [19]: 1) Initial exposure to oxygen plasma results in a rapid general oxidation of the surface, producing hydroxyl, carbonyl and hydroperoxide groups capable of strong polar interaction with the epoxy resin. Contact angle measurements show changes in surface energy, and the liquid epoxy resin (before curing) can wet the surface of the fibres. 2) Cross-linking of the polyethylene surface by the plasma UV radiation improves adhesion by increasing the cohesive strength, forming a tough skin (~0.1 μm thickness). 3) Longer treatment times cause extensive pitting of the fibre surface, improving adhesion by mechanical keying. On the basis of these fundamental studies on monofilaments, a continuous process has been developed for plasma treatment of polyethylene fibres and the adhesion of these fibres has been evaluated in composites, using short beam shear tests. For undirectional fibre composites containing 60% fibre the interlaminar shear strengths measured in this way are increased from 15 to 30 MPa which is close to the shear strength of oriented polyethylene.

Although the major applications of polyethylene fibres are in ropes, cords, protective clothing etc., substantial effort has been given to their possible application in composites (e.g. crash helmets). Major advantages stem from the high ductility of the fibres which leads to high energy absorption and high damage tolerance. Hybrid composites, where polyethylene fibres are incorporated into resin matrices in conjunction with glass or carbon fibres, are especially promising [20] and have led to applications in surf-boards and skis respectively.

A recent development has been the production of 100% polyethylene fibre composites by a hot compaction procedure [21] where the fibres are compacted closely and by melting a small fraction, solid sections with a very high degree of integrity are produced. Biomedical applications for these solid materials are envisaged, for example, orthodontic brackets or bone plates.

## Ram extrusion, hydrostatic extrusion and die drawing

Ram extrusion and hydrostatic extrusion have been explored for the production of highly oriented polymers, especially polyethylene, since the early 1970s by Porter and his collaborators [22] in the USA, Takayanagi and his collaborators [23] in Japan, and at Leeds University [24, 25] in the UK. Progress has been achieved in two respects: 1) materials with very high stiffness and strength have been produced, especially by combining extrusion with further tensile drawing, and 2) an understanding of the mechanics of solid phase processing has developed.

However, it has become apparent, that except perhaps for exceptional applications, where cost is not a major criterion, ram and hydrostatic extrusion have the disadvantages of being rather slow batch processes. This recognition led to the invention of the Leeds die-drawing process [26], where rod and sheet are produced by drawing through heated dies, and, by incorporation of a mandrel into the die, biaxially oriented tube. In these solid phase deformation processes, it can be considered that the deforming polymer follows a given route across the stress-strain-strain rate surface [25]. As shown in Fig. 6, the processing route for die-drawing is much more favourable than that for hydrostatic extrusion where the highest strain rates coincide with the highest degrees of plastic deformation at the die exit, so that in the latter case the flow stresses are very large.

The die drawing process has been found to be applicable to a wide range of both crystalline and amorphous polymers, including polyethylene,

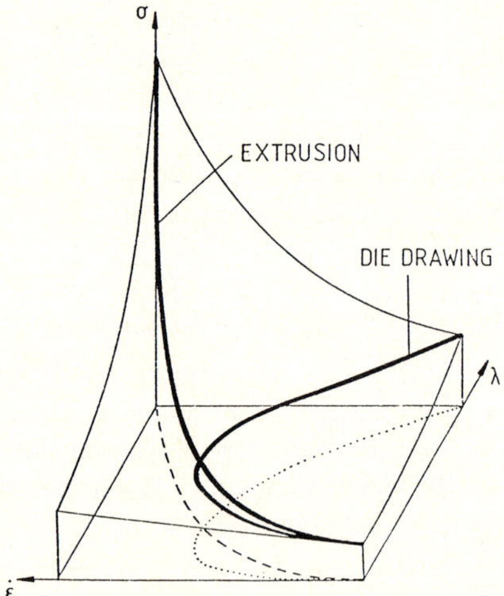


Fig. 6. Schematic extrusion and die drawing paths across true stress-draw ratio-strain rate surface [equal production speeds, same die]

polypropylene, polyethylene terephthalate, polyvinylchloride and polycarbonate. From this exploration of the potential of the die-drawing process, several key points have emerged:

1) Valuable enhancement of properties can be obtained without necessarily achieving the levels of modulus and orientation required for fibres or monofilaments. Improvements in strength and barrier properties are often as important or more important than dramatic increases in stiffness, which may even be a disadvantage.
2) For many solid section applications, biaxial orientation is needed, and for pipes, hoop stress is of particular importance.
3) For commercial exploitation, a viable continuous process is required with production speeds of at least metres/minute.

Recent research at Leeds has been concerned with an analysis of the mechanics of the die-drawing process [27]. This has followed similar lines to that adopted previously for hydrostatic extrusion. The starting point for the analysis is the Hoffman-Sachs lower bound solution where a force balance is obtained for deforming elements through the die. Because the flow stress of a polymer is very dependent on both strain and strain-rate, it is necessary to obtain independent information on these issues, i.e. to determine the stress-strain-strain rate surface. The assumption that it is adequate to use such a surface to predict the flow stress at any point in the die is an essential ingredient of the Leeds approach. It implies that there is no history dependence and that the flow stress is dependent only on the relevant strain rate at the particular point in the die and the total plastic strain imposed, irrespective of the deformation path chosen. It appears from comparison of the results of the analysis with experimental data that this is a valid assumption.

A complication in the die-drawing process is the possibility of adiabatic heating due to the heat produced by plastic deformation [28]. It is shown that this is appreciable, but that there is also significant conduction of heat through the polymer in both the axial and radial directions. The effect of the temperature rises can be monitored by measuring the drawing stresses, which were found to be in good agreement with those predicted, for a range of drawing speeds.

As indicated, an attraction of die-drawing is the possibility of continuous processes, where melt extrusion of the polymer is followed in-line by die-drawing to the final product. Particular effects have been directed to the continuous production of biaxially oriented polyethylene pipes, which are suitable for gas or water distribution. A schematic diagram of the Leeds biaxial die drawing facility is shown in Fig. 7. The start up of this continuous process for pipes required the development of a novel procedure. First, a length of tube is belled at one end using a belling mandrel and rod. The belled end can then be fitted over the mandrel in the die and the other end is joined to the extruded undrawn tube from the melt extruder by butt fusion. At the start-up the belled end is heated to 120—125 °C and drawn over the mandrel by the second stage caterpillar to produce biaxially oriented tube.

In the initial studies of the die-drawing process, the physical property measurements were principally concerned with stiffness. It was soon established that, for uniaxially oriented products, the modulus related to the draw ratio, provided that the die was maintained at temperatures at least 10—20 °C below the melting point of the polymer. In polyethylene, polypropylene and polyoxymethylene it was found possible to achieve modulus levels comparable to those previously obtained by tensile drawing of filaments and tapes, or hydrostatic extrusion of rods.

The next step was to produce oriented sheet and tube. Oriented sheets of several polymers have

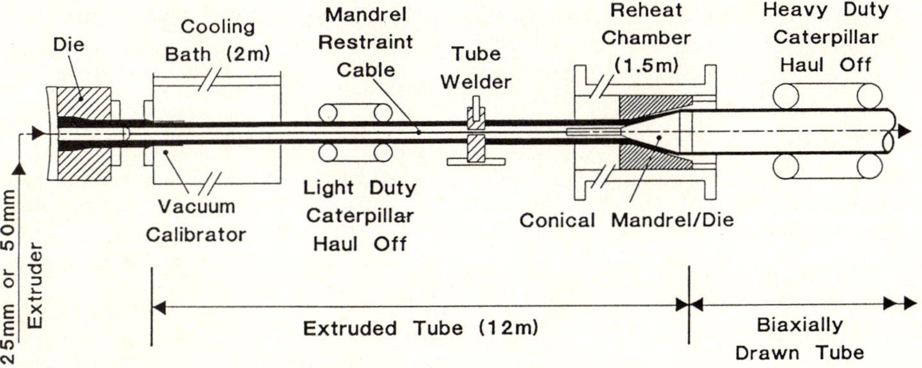


Fig. 7. Continuous die drawing line for biaxially drawn tube

been made by extrusion through a slit die, under conditions where the lateral width is almost constant (plane strain deformation). This produces biaxially oriented sheets, and in polyethylene a particular study has been made of their fracture toughness, using a special facility where the tests are undertaken under a substantial hydrostatic pressure (~500 MPa). This hydrostatic pressure increases the yield stress of the polymer and ensures that brittle fracture occur. The results for specially selected grades of polyethylene (certain copolymers or polymers with a bimodal molecular weight distribution) are remarkable in showing that the fracture toughness is substantially increased by this process, even for the propagation of cracks parallel to the principal orientation direction [29].

In the case of pipes, it has been particularly important to increase the hoop strength by imposing a positive deformation in the hoop direction by expanding the tube over a mandrel as shown in Fig. 7. It is of special interest that such biaxially oriented tubes always fail in a ductile manner, even in long-term tests. Moreover, their stress rupture behaviour shows very significant improvement over isotropic pipis of the same polymer (Fig. 8). Another aspect of the Leeds research on die-drawn pipes has been to establish satisfactory procedures for joining the pipis by butt fusion and electrofusion socket welding. Electrofusion coupling is an excellent method of joining biaxially oriented pipes. Although the outer skin of the pipe loses its orientation, this is not serious because the hoop orientation is lowest in this layer, so that the pipe retains almost all the strength and stiffness associated with the biaxial orientation.

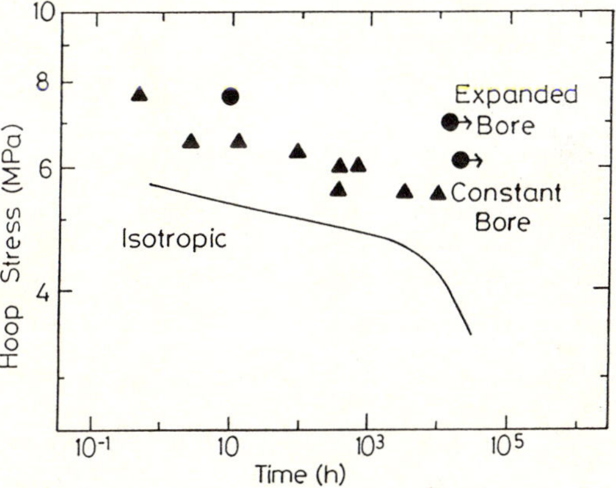


Fig. 8. Stress rupture behaviour of PE pipes at 80 °C. ▲ die drawn constant bore; ● die drawn biaxially oriented expanded bore; ●→ not yet failed

Butt fusion is also remarkably successful, in that at first sight one would expect to lose substantial amounts of molecular orientation in the melted region where the join occurs. However, the weld zone can be made sufficiently narrow that weld strengths reaching 40% of the unwelded biaxially oriented pipe are obtained, which is still substantially greater than the strength of isotropic pipe.

## Conclusions

Substantial progress is being maintained with regard to both the development and exploitation of high modulus and high strength flexible polymers.

Polyethylene fibres are now complementary to carbon and Kevlar fibres in a wide range of applications, partly as a result of the successful development of enabling inventions such as plasma surface treatment. The new die-drawing process has now been developed for the continuous production of biaxially oriented pipes for gas and water distribution and monofilaments for ropes and cables. It is of particular interest that a range of biomedical applications are being actively considered for both polyethylene fibres and the die-drawn products.

## References


1. Andrews JM, Ward IM (1970) J Mater Sci 5:411
2. Ward IM (1962) Proc Phys Soc 80:1176
3. Ward IM (1971) Mechanical Properties of Solid Polymers, John Wiley & Sons Ltd, London, Chapter 10
4. Treloar LRG (1960) Polymer 1:95, 279, 290
5. Sakurada I, Ito T, Nakumae K (1960) J Polym Sci C15:75
6. Capaccio G, Ward IM (1973) Nature Physical Science 243:143. Brit patent Appl 10746/73 (filed 3 March 1973)
7. Zwijnenburg A, Pennings AJ (1976) J Polym Sci (Letters) 14:339
8. Smith P, Lemstra PJ, Kalb B, Pennings AJ (1979) Polym Bull 1:733
9. Smith P, Lemstra PJ (1980) J Mater Sci 15:505
10. Amornsakchi T, Jawad SM, Cansfield DLM, Pollard G, Ward IM (1993) J Mater Sci 28:1689
11. Flory PJ (1945) J Am Chem Soc 67:2048
12. Termonia Y, Greene WR, Smith P (1986) Polymer Comm 27:295
13. Wilding MA, Ward IM (1981) Polymer 22:870
14. Sherby OD, Dorn JE (1956) J Mech Phys Solids 6:145
15. Mercer FB (1986) Philips Lecture, Royal Society, 9th October
16. Woods DW, Busfield WK, Ward IM (1985) Plastics and Rubber Proc and Appl 5:157
17. Jones R, Salmon A, Ward IM (1993) J Polym Sci Polym Phys (ed) 31:807
18. Ladizesky NH, Ward IM (1983) J Mater Sci 18:533
19. Tissington B, Pollard G, Ward IM (1992) Composites Science and Technology 44:185
20. Ladizesky NH, Ward IM (1986) Composites Science and Technology 26:199
21. Ward IM, Hine PJ, Norris K (1992) Brit Patent Appl No 9204965.9 (filed March 1992)
22. Weeks NE, Porter RS (1974) J Polym Sci Polym Phys Ed 12:635
23. Nakmura K, Imada K, Takayanagi M (1972) Int J Polym Mat 2:71
24. Gibson AG, Ward IM, Parsons B, Cole BN (1974) J Mater Sci 9:1193
25. Coates PD, Gibson AG, Ward IM (1980) J Mater Sci 15:359
26. Coates PD, Ward IM (1979) Polymer 20:1553
27. Motashar FA, Unwin AP, Craggs G, Ward IM, Polym Eng and Sci (in press)
28. Kukureka SN, Craggs G, Ward IM (1992) J Mater Sci 27:3379
29. Tsui S-W, Duckett RA, Ward IM (1992) J Mater Sci 27:2799


Author's address:

Prof. Dr. I. M. Ward
Interdisciplinary Research Centre
for Polymer Science and Technology
University of Leeds
Leeds LS2 9JT, United Kingdom

**Progress in Colloid & Polymer Science** Progr Colloid Polym Sci 92:111—119 (1993)

# Lamellar morphology of polydiacetylene thin films and its correlation with chain lengths

C. Albrecht*), G. Lieser, and G. Wegner

Max-Planck-Institut für Polymerforschung, Mainz, FRG

*Abstract:* Thin films of the soluble polydiacetylene derivative P-4-BCMU were prepared from chloroform solution. During the evaporation of the solvent the solution becomes anisotropic before the solid state is reached. The morphology of these films is lamellar, the lamellae being seen edge-on. Lamella thickness is proportional to the chain length of the worm-like polymers. Formation of lamellae is a function of the dwell time of the polymer in the liquid crystalline state. In the early stages lamella thickness is determined by the average number molecular weight. Chain ends are concentrated in disordered interlamellar regions. Approaching equilibrium chain segregation with respect to chain lengths is observed. In addition, the effect of electron irradiation damage in the polymer films is investigated.

*Key words:* P-4-BCMU — main chain LCP — correlation of lamella thickness with $M_n$ — irradiation damage

## Introduction

Polydiacetylenes are linear polymers with conjugated backbone. The system of conjugated multiple bonds gives the chain worm-like conformation and is responsible for the intense color of the macromolecules. The polymers are synthesized via a topochemical reaction [1] of a large variety of diacetylene derivatives in the crystalline state of the monomers. Only a few of the polymers are soluble [1—3] in organic solvents.

In this work, we characterize in particular the solid state of one of the soluble derivatives. It is the poly(5,7-dodecadyine-1,12-diol bis(((4-butoxycarbonyl)methyl)urethane)), henceforth referred to as P-4-BCMU. Scheme 1 exhibits the repeating unit of the polymer. Flexible side chains are attached to the worm-like chain. They have the pecularity that they can interact with each other via hydrogen bonds. We showed in previous publications [4—6] that during solidification from a dilute solution the polymer passes a liquid crystalline state under the formation of lamellae. Lamellar morphology is a very familiar picture in the case of monodisperse smectic low molecular weight liquid crystals. This should at first sight not be different for worm-like or rod-like polymers as long as the aspect ratio exceeds a critical value. Indeed, Frenkel [7] showed in a molecular dynamics calculation lamella formationin an anisotropic solution of spherocylinders of an aspect ratio of 5 as a function of polymer concentration. The model of monodisperse chains used, however, is unrealistic because polydispersity is a dominant feature of polymers. They are therefore thought to form only nematic liquid crystals.

Even so, the thickness of the observed lamellae in the order of magnitude of 100 nm gives rise to the

Scheme 1

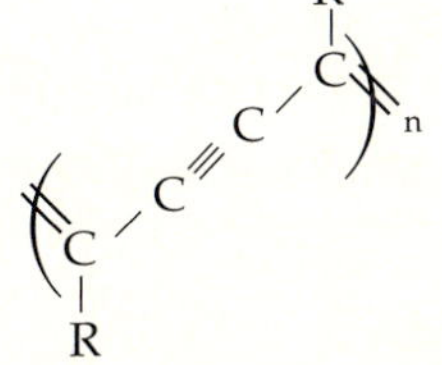

$$\left( \underset{R}{\underset{|}{C}} - C \equiv C - \overset{R}{\overset{|}{C}} \right)_n$$

$$R = -(CH_2)_4-O-\overset{O}{\overset{\|}{C}}-\underset{H}{\underset{|}{N}}-CH_2-\overset{O}{\overset{\|}{C}}-O-C_4H_9$$

*) Present address: Hoechst AG, 65203 Wiesbaden, FRG.

assumption that it is controlled by the chain length. In this investigation, we want to establish the correlation between the average molecular weight of the polymer, its molecular weight distribution, and lamella thickness and shape.

## Experimental

### *Synthesis*

Oxidative coupling of 1-hexinol (**1**) in the presence of twovalent copper compounds in pyridine yields dodeca-5,7-dyine-1,12-diol (**2**) according to Eglinton and McCrae [8] (Scheme 2). Diacetylene (**2**) is reacted with butylisocyanoacetate to the butoxycarbonylmethylurethane of dodeca-5,7-dyine-1,12-diol (**3**), abreviated as 4-BCMU. The monomer is well crystallizable to lamellar crystals which were irradiated in a $^{60}$Co-γ-source at a temperature of 0°C till a dose of 300 MGray was accumulated. The molecular weight reached under these conditions a value of $\overline{M_w}$ = 555 600. Beside this preparation, we also used a sample polymerized at room temperature with lower molecular weight and broader molecular weight distribution. The bronze-colored P-4-BCMU crystals were dissolved in chloroform and then precipitated in methanol in order to remove residual monomer and low oligomers.

Both samples were used as the highest in two series with decreasing molecular weights produced from the pristine species by ultrasonic degradation in a Bandelin Sonorex RK 514H device. Dilute P-4-BCMU solutions in chloroform ($c$ = 1 g/l) were exposed to ultrasonication in 20 ml flasks up to 1 h.

### *Characterization*

Molecular weights and molecular weight distributions were determined by means of gel permeation chromatography (GPC). Weight average molecular weights of four samples were determined by light scattering in an ALV light scattering apparatus. These samples were used for calibration of GPC curves. Light source was a krypton laser with a wavelength of 647.1 nm. Each polymer sample was dissolved to four different concentrations between 2 and 6 g/l in dimethylformamide (DMF). The solutions were purified by filtering through a 0.5 μm Millipore filter. Concentrations were determined after filtration by UV/VIS spectroscopy in a Perkin-Elmer Lambda 3 spectrometer. Calibration was performed by measurement of the extinction of P-4-BCMU solutions at 470 nm as a function of well known concentrations. The refractive index increment for P-4-BCMU in DMF is in $dn/dc$ = 0.1467 ml/g.

For GPC analysis a Waters model 590 apparatus was used with Waters software 480. Detection was performed at 460 nm with a SOMA UV/VIS detector. Solvent was stabilized THF.

Thin films for observation in a transmission electron microscope (Zeiss EM 902, run at a high voltage of 80 kV) were prepared by casting a dilute solution of P-4-BCMU in chloroform (concentration 1 g/l) onto glass slides. In most cases the evaporation of the solvent was retarded by covering the slide with a watch glass. Lamella formation is subject to the evpoaration rate, i.e., to the dwell time in the state of anisotropic solution. Retardation by covering the solution with a watch glass is sufficient for the lamellae to become visible but too short to attain their equilibrium shape. After evaporation of the solvent the films were floated off the glas onto

Scheme 2

$$2\ HO{-}(CH_2)_4{-}C{\equiv}CH \xrightarrow{Cu^{2+}/pyridine} HO{-}(CH_2)_4{-}C{\equiv}C{-}C{\equiv}C{-}(CH_2)_4{-}OH$$

(**1**) (**2**)

$$HO{-}(CH_2)_4{-}C{\equiv}C{-}C{\equiv}{-}(CH_2)_4{-}OH + 2\ O{=}C{=}N{-}CH_2{-}COOC_4H_9 \longrightarrow$$

(**2**)

$$(C_4H_9OOC{-}CH_2{-}OCO{-}NH{-}(CH_2)_4{-}C{=}C{-})_2$$

(**3**)

a water surface and transferred to hexagonal 600 mesh copper grids. The films are selfsupporting and can be examined without further preparation in a transmission electron microscope. Replicas were prepared by shadowing the films on glass slides by simultaneous evaporation of platinum and carbon and subsequent carbon coating. After floating off and transfer of the shadowed films the coated grids were immersed into chloroform to dissolve the polymer. Electron diffraction patterns were recorded in a Philips EM 300 electron microscope at a sample temperature of 150 K.

Average spacings between lamellae were determined by application of optical transform to negatives of electron micrographs in suitable electron optical magnifications. The optical diffraction patterns were registrated on photographic film. Light source for the device was a 5 mW HeNe laser. The diffraction length was calibrated by an optical grid with 100 lines/cm.

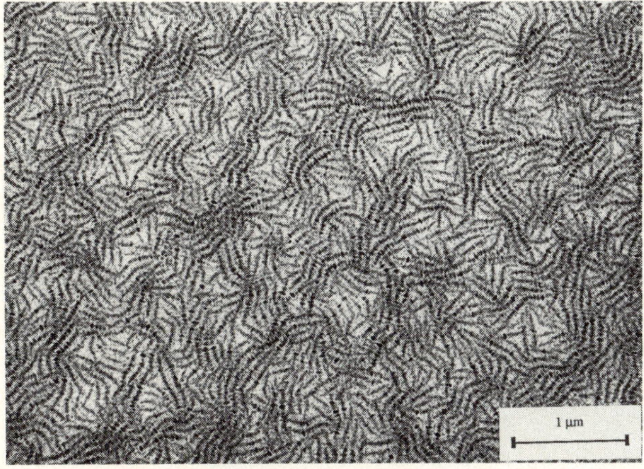


Fig. 1. Solid P-4-BCMU film from a 1 g/l solution in chloroform, micrograph is superposed by contamination

## Results and discussion

Electron microscopical observation of solution cast thin polymer films is a frequent first approach to study polymer morphology. In the very beginning of this investigation thin P-4-BCMU films were prepared by pouring a chloroform solution of the polymer onto a glass slide, allowing the solvent to evaporate under ambient conditions. The evaporation is usually completed after a few minutes. Figure 1 displays a typical transmission electron micrograph of such as P-4-BCMU film prepared without special care. Instead of a homogeneous film as in the case of many conventional polymers it was surprising to see numerous ribbons with an average width in the order of magnitude of 100 nm extended over the whole field of view. The ribbons display striations originating from surface roughness. The contrast of the striation is enhanced with time of exposure to the electron beam by inhomogeneous contamination by residual hydrocarbons from the vacuum system of the microscope. Surface roughness offers seeds to the precipitation which is finally transformed into carbon by the electron beam. One can demonstrate that the striations label the chain direction [4, 6]. For the registration of electron diffraction patterns the primary beam intensity has to be lowered by at least two orders of magnitude with respect to imaging conditions for Fig. 1. When it is possible to find small areas in the sample where a few bands are nearly parallel to each other an electron diffraction pattern looks like the one displayed in Fig. 2. The intensity is distributed continuously along lines the position of which is identical with the layer lines in an x-ray fiber diagram of the drawn polymer indicating that the chains are oriented parallel to the substrate. The observed bands are therefore interpreted as lamellae seen edge-on. Within the lamellae adjacent chains are not in register. This essential feature of a liquid crystalline structure is maintained in the solid state although the solvent has evaporated.

In view of polymer morphology, Fig. 1 is best comparable to electron micrographs of ultrathin sections of stained partially crystalline polyolefins where lamellae are also seen edge-on [9, 10]. Their thickness, however, does not esceed several 10 nm. Lamella thicknesses in the order of 100 nm are known from extended-chain crystals [11—13] which can be prepared under high hydrostatic pressure or which appear directly in a process of simultaneous polymerization and crystallization. Other examples of lamella formation are glassy block copolymers in a distinct ratio of block lengths [14]. But none of these examples match the case of P-4-BCMU because, on one hand, its chains are worm-like and unable to fold and, on the other hand, because the formation of thick lamellae occurs under ambient conditions.

In order to approach a solution of the question about the magnitude of the lamellae thickness, we

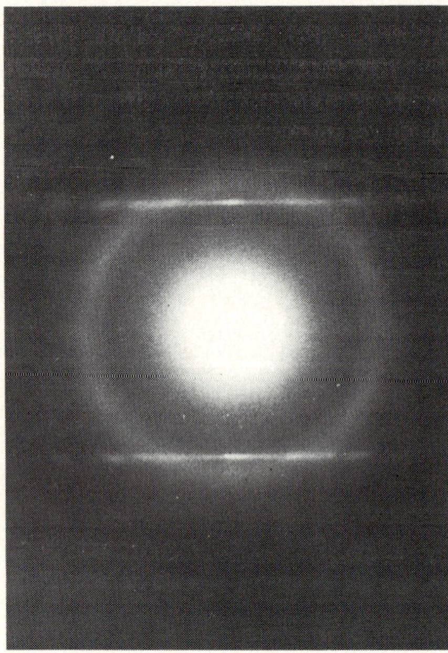

Fig. 2. Electron diffraction pattern of an area covering only a few lamellae in the same orientation

studied the morphology of a series of samples with various chain lengths. Two primary samples which had undergone varying polymerization conditions were available. Both were degraded by ultrasonication to obtain a series of polymers with various molecular weights. In contrast to photodegradation [15] ultrasonication [16—18] has the advantage of better reproducibility. In addition larger quantities can be handled at a time. Ultrasonication disrupts long chains by mechanical shear which is produced by cavitation inside a solution. Degradation is terminated at a lower limiting degree of polymerization which is determined by the power of the ultrasonic source used. At this limit the chains are so short and mobile that they are able to orient instantaneously along sudden changes of the shear field.

The molecular weights of the series of samples were determined by GPC after calibration as described in the experimental section. Figure 3 shows a linear relation between the retention times at the positions of the maxima of GPC curves and the average molecular weights measured by light scattering. We used the same calibration for all samples ignoring the different polydispersity of both primary samples. For the sake of comparability of lamella thickness and chain length the determ-

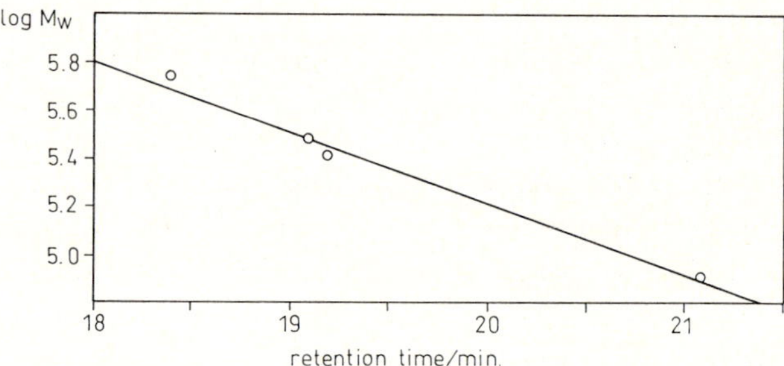


Fig. 3. Calibration of GPC with samples of known $\overline{M_w}$

ination of $\overline{M_n}$ was necessary for all samples. It was necessary for all samples. It was derived by subdividing the calibrated GPC curves into narrow strips corresponding to a sequence of retention times and treating the respective intensities as the frequencies of monodisperse polymers. In Table 1 the molecular weights of samples from which films for electron microscopic examination were prepared are listed. It is worth to note that the polydisperity of the degraded samples has not significantly changed with respect to the pristine samples indicating that the degradation is statistical.

Table 1. Molecular weights and polydispersity of P-4-BCMU — samples prepared by ultrasonic degradation

| Series | $\overline{M_w}$ | $\overline{M_n}$ | $\overline{DP_n}$ | Polydispersity |
|---|---|---|---|---|
| 1 | 515 000 | 216 400 | 425 | 2.4 |
| 1 | 366 100 | 138 000 | 271 | 2.7 |
| 1 | 89 500 | 38 900 | 76 | 2.3 |
| 2 | 350 200 | 63 500 | 125 | 5.5 |
| 2 | 362 000 | 50 200 | 99 | 7.2 |
| 2 | 274 100 | 27 900 | 55 | 9.8 |
| 2 | 158 400 | 19 200 | 38 | 8.3 |
| 2 | 119 400 | 15 800 | 31 | 7.6 |

Figure 4 shows electron micrographs of P-4-BCMU films for various molecular weights. In contrast to the preparation of the film shown in Fig. 1 solvent evaporation was retarded and film thickness is reduced. It is obvious that the lamella thickness is the higher the longer the molecules are. The thickness of the lamellar core was measured on magnified prints and the respective data are presented as a function of the chain length calculated from the respective number average molecular weights (squares in Fig. 5). In addition

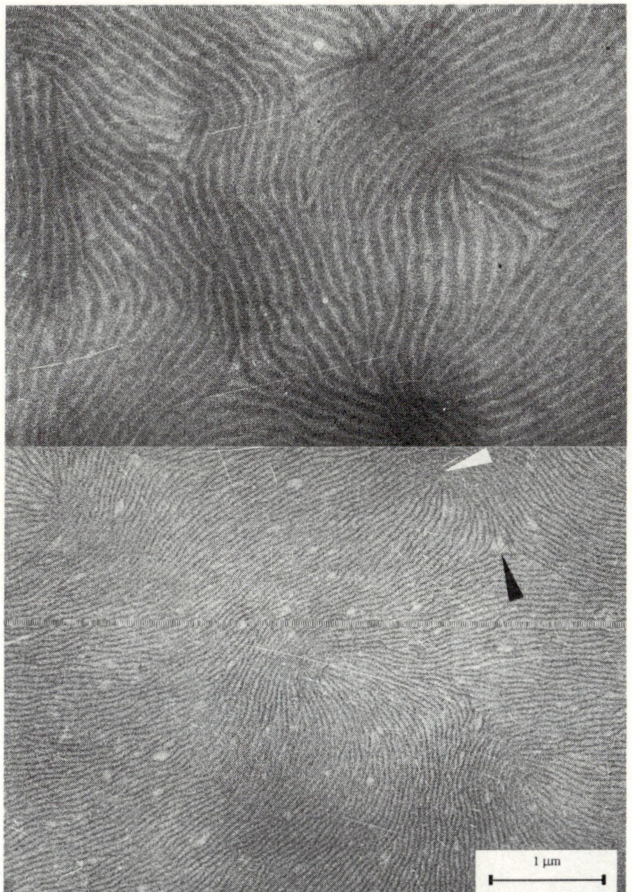

Fig. 4. Solid P-4-BCMU films of samples degraded by ultrasonication: a) after 5 min, $\overline{M_n}$ = 64 000 b) after 45 min exposure time, $\overline{M_n}$ = 16 000. The arrows in (b) indicate two disclinations with $s = +1/2$ (white) and $s = -1/2$ (black)

to the data a straight line of slope 1 is drawn which fits the values very well. On the other hand, there are gaps of increasing size between lamellae when the chains are short. This is expressed by increasing deviations between the long periods (triangles in Fig. 5) and the straight line. Reasons for this behavior are mainly kinetical. This aspect will be discussed later. Independent from variations of lamella thickness both micrographs of Fig. 4 show the morphology of a frozen liquid crystalline structure with disclinations of strengths $s = \pm 1/2$ as its typical singularities. Two of them are labelled in Fig. 4b. The aspect of liquid crystallinity in this type of systems is treated elsewhere [6].

By the experimental results were arrive at the conclusion that laemlla formation is not an exclusive

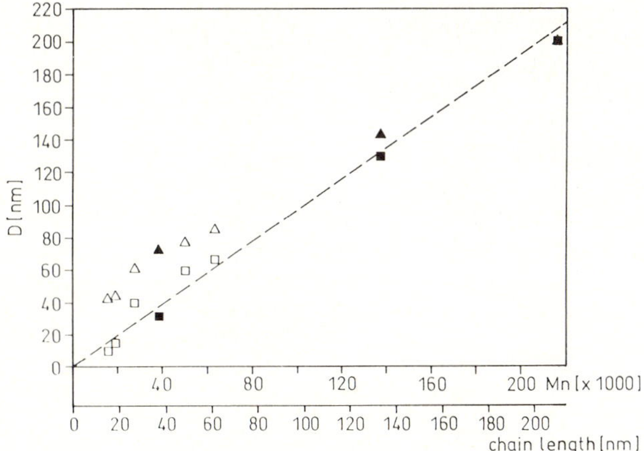


Fig. 5. Correlation of lamella thickness (□) and long period (△) with chain length. The filled symbols belong to samples originating from degradation of a pristine sample with higher molecular weight and more narrow molecular weight distribution

porperty of monodisperse molecules but is a possible process of organization in main chain polymer liquid crystals too. Lamella thickness is essentially determined by the number average degree of polymerization. When the transition from isotropic to anisotropic solution has taken place and the macromolecules have oriented parallel to each other it is obviously favorable for the system to concentrate as many chain ends as possible in interlamellar regions. This is expressed by the fact that $\overline{M_n}$ is determining the lamella thickness. Shorter chains are included inside the lamellae, longer ones act as links between lamellae. As a matter of fact, the number average molecular weight can be measured from electron micrographs even if the high molecular tail of the molecular weight distribution is concealed.

In order to get an idea of the respective amount of chains the length of which exceeds the average, we suppose a Schulz-Flory most probable distribution, ignoring that in our case the macromolecules are more broadly distributed. The frequency distribution is then given by

$$F(p) = \alpha^{p-1}(1 - \alpha) , \tag{1}$$

$\alpha$ being the extent of reaction in the case of a polycondensation and $p$ the degree of polymerization. With the approximation $-\log \alpha = 1 - \alpha$ which is valid provided that $(1 - \alpha)^2 \ll (1 - \alpha)$ we get rid of the parameter $\alpha$. With $P_n = 1/(1 - \alpha)$, we can write

$$F(p) = \frac{1-a}{a} a^p = \frac{1}{aP_n} \exp(p \cdot \log a)$$

$$= \frac{1}{P_n} \exp(-p/P_n) , \tag{2}$$

because $a \lesssim 1$. The cumulative number fraction is given by

$$\int_0^P F(p)dp = (1/P_n) \int_0^P \exp(-p/P_n)dp$$

$$= -\exp(-P/P_n) + 1 . \tag{3}$$

We are interested in the fraction of molecules $R(P)$ the degree of polymerization of which exceeds the value $P_n$, i.e., the fraction under the tail of the molecular weight distribution. For this fraction $R(P)$, we get simply

$$R(P) = 1 - \int_0^P F(p)dp = \exp(-P/P_n) . \tag{4}$$

For a most probable distribution 37% of the chains exceed $P_n$ and 14% are longer than $2P_n$. Taking into regard that our samples are more broadly distributed, we see that more than 14% of the chains are incorporated in at least two subsequent lamellae bridging the disordered interlamellar regions and preventing the film from disintegration.

At this point it has to be mentioned that the polymer is sustaining irradiation damage during electron microscopical observation. This is best demonstrated by two micrographs of a film which was prepared without any special care concerning evaporation rate of the solvent. Both micrographs in Fig. 6 were exposed at 195 K at low beam intensity. Figure 6a shows a structureless film not regarding variations of mass thickness. After exposure of the micrograph the film was irradiated in its center with focussed condenser for several minutes. A subsequent exposure — still at the low temperature — shows the familiar lamellar morphology (Fig. 6b). The initial absence of morphological details is confirmed by means of replicating techniques. Replicas of pristine films which were prepared at high evaporation rate of the solvent (short dwell time in the state of anisotropic solution) are entirely uniform.

Nevertheless, in this stage a difference between lamella cores and interlamellar regions is already existing, even though accumulation of chain ends is still in progress. Irradiation damage is selective and affects the disordered interlamellar region more than the lamella core. It is, however, not possible to burn holes in the interlamellar regions by continuous irradiation. We suggest therefore that the mechanism of disintegration proceeds from chain ends in disordered surrounding toward the surface of the lamellar core. In this particular case irradiation damage acts as a selective stain.

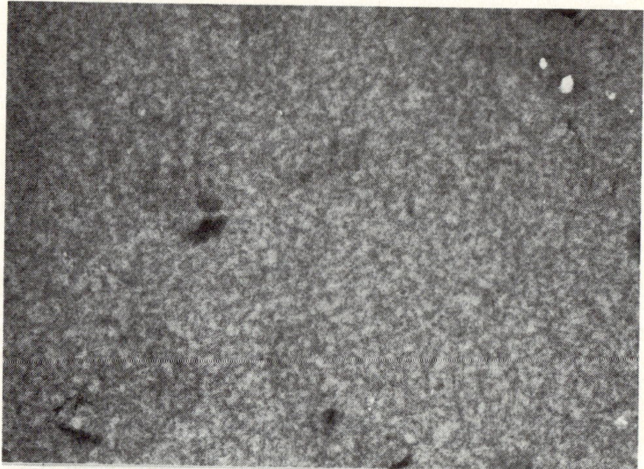

a)

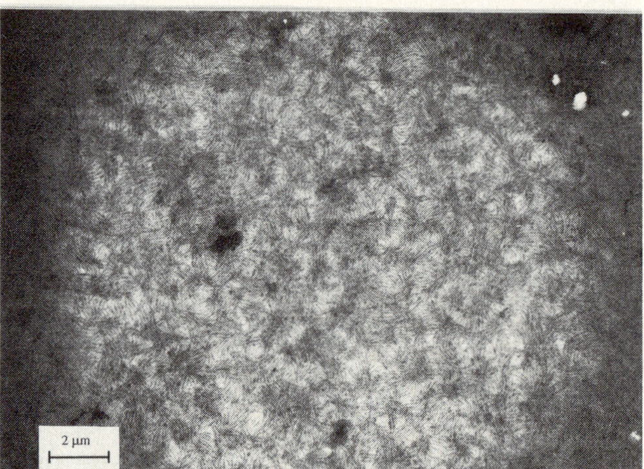


b)

Fig. 6. Electron micrographs of a solid P-4-BCMU film, recorded at 195 K at low beam intensity: a) immediately after the sample was hit by the beam, b) same area after the sample was irradiated in the center by the focussed beam

In order to check the electron microscopical observation that chain cleavage is more probable near an isolated chain end projecting into the disordered region the effect of irradiation on the molecular weight was also studied in the bulk. In a separate experiment a 80 μm thick film of the broader distributed primary sample was exposed to electrons inside a van de Graaff accelerator until a dose of 100 kGray was accumulated. The converted quantity of polymer was sufficient for molecular weight determination. In contrast to the conditions of a thin film in the electron microscope the cleaved chain fragments inside the sample are not volatile enough to evaporate. The sample was therefore again extracted by methanol to eliminate the cleaved short chains. The irradiated polymer was fully soluble in chloroform, indicating that no cross-linking had taken place. GPC analysis of the polymer after extraction showed that the average molecular weight was only insignificantly reduced, but the polydispersity had diminished from 7 to 4. This result shows that the destructive interaction with a high energy electron beam is largely restricted to chain ends in disordered surrounding. Chains passing the interlamellar gap are not influenced because otherwise the average molecular weight would have shifted to significantly lower values.

Two things, however, should not be confounded: lamella formation is not originated by selective irradiation damage. It is a convenient means to make lamellae visible in the early stages of an ordering process which proceeds with distinct kinetics in an anisotropic solution. In this stage the electron density difference is not yet sufficient to produce contrast in an electron microscopic image and by this reason pristine films look uniform. Also SAXS patterns of thick samples do not show longperiods in the early stage of lamella development.

It would have been too time-consuming to elaborate kinetics for all samples, all the more so as kinetics vary with molecular weight. Therefore, the films used in the present investigation underwent different kinetics with respect to the molecular weight and they solidified at varying distances from equilibrium. For low molecular weights the critical concentration is high, the concentration range up to solidification is narrow [6], and therefore the dwell time in the anisotropic solution is shorter at constant evaporation rate. We suggest that in samples with lower molecular weight an earlier state of lamella formation was frozen in which the interlamellar regions are broader.

When the ordering process proceeds further we observe by replication technique that gaps beween lamellae are also present in unirradiated samples. Figure 7 displays a replica of a P-4-BCMU film which was allowed to stay in the state of an anisotropic solution for a long duration of time. We suggest that the lines transverse to lamellae occur as a consequence of the volume shrink during the solidification when the solvent evaporates suddenly. They are oriented parallel to the molecular axes and label the director field of the liquid crystal [6]. Beam damage can be ignored the more so the lamella surfaces are flat; it may sharpen the contour of lamellae, but is then not necessary for their visibility.

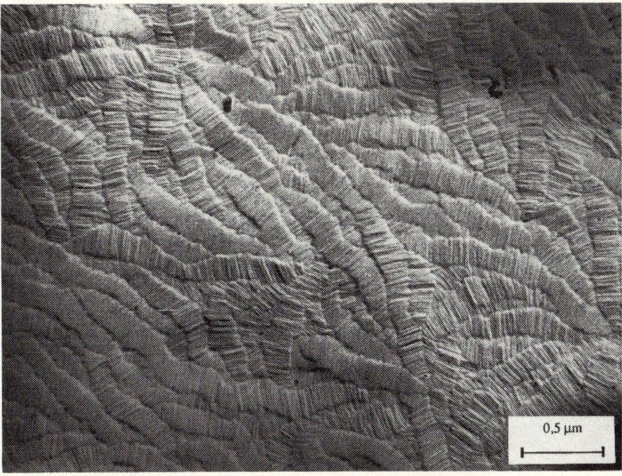


Fig. 7. Replica of a pristine solid P-4-BCMU film which was in the liquid crystalline state for a long duration of time

When the liquid crystalline state is maintained for a long duration of time (on the order of 2 weeks) the system approaches equilibrium by further separation with respect to the chain length. This process is driven by pairwise annihilation of disclinatiaons (cf. Figs. 4 and 8a) and accompanied by local changes of orientation and formation of new lamellae [6]. Simultaneously, short chains are repelled from the interior of the lamella core. They migrate either across the chain axes to the ends of lamellae giving them a pointed shape (Fig. 8a) or they move along the chain direction and assemble in thin lamellae between thick ones. Such an alternating sequence of thick and thin lamellae is displayed in Fig. 8b. At this late stage near equilibrium lamella shape more reflects the molecular weight distribution than the number average molecular weight. In addition, both mechanisms lead to smoothing of lamella surfaces and minimize the volume of disordered regions. The wide gaps between lamellae observed in Fig. 4b which are responsible for the discrepancy between

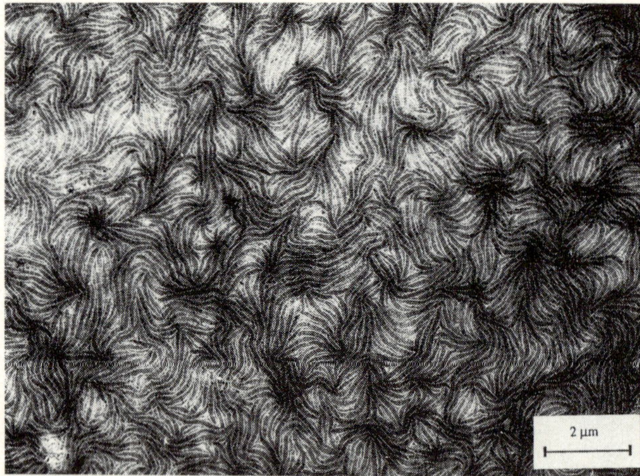


a)

b)

Fig. 8. Solid P-4-BCMU film after exlucision of short chains from thick lamellae: a) short chains at the outer tips of lamellae, b) formation of additional thin lamellae

lamella thickness and longperiod obvious in Fig. 5 are closed in the near-equilibrium state.

## Conclusions

Worm-like macromolecules like P-4-BCMU take lamellar morphology when they are able to form a mesophase and to maintain it for some duration of time. Obviously, Gibbs' free energy is lowered when distortions like chain ends can be concentrated in relatively small volumina. This ordering principle can also be observed at partial crystallization of some flexible copolymers. Similar morphology is also known from semiflexible thermotropic polyesters [19—22]. The important difference to these systems in which the chains are able to fold back is the fact that, in the case of main chain liquid crystalline polymers, the lamella thickness is determined by the chain length. The lyotropic liquid crystalline state grants sufficient mobility to the chains to facilitate total segregation with respect to their lengths. This process seems to be general for rigid macromolecules although it is not yet observed in thermotropic main chain liquid crystalline polymers. The viscosity in these systems might be too high to prevent chain segregation within reasonable times. Derivatives of phthalocyaninatopolysiloxane are other lyotropic systems where lamella formation is observed [5] on a smaller scale because the chains are much shorter. In bilayers of one of these polymers prepared in a Langmuir trough it could be confirmed by high resolution electron microscopy [23] that lamella thickness coincides with chain length, too, but segregation of chains of various lengths has not taken place. In the double layer also the long tail of the molecular weight distribution is seen. In the best studied lyotropic polymer system, poly(-p-phenylene terephthalamide) [24], similar observations are not reported. The system is possibly too viscous and the time during which the aligned liquid crystalline state is maintained in the usual wet-spinning process is very short. In addition, electron microscopic investigations of this aromatic polyamide are not easy. According to this point of view, P-4-BCMU is certainly a very favorable case where, in addition, the chains are long enough to be unpretentious with respect to microscopic resolution. The result that chains are able to segregate in a liquid crystal polymer with respect to their length is important and has to be taken into regard when physical properties of polymers with LC history are discussed.

## References


1. Wegner G (1969) Z Naturforschung 24b:824
2. Patel GN (1978) Polym Preprint. Am Chem Soc Polym Chem Div 19:160
3. Wenz G, Müller MA, Schmidt M, Wegner G (1984) Macromolecules 17:837
4. Albrecht C, Lieser G, Wegner G (1989) Beitr Elektronenmikroskop. Direktabb Oberfl (BEDO) 22:351
5. Lieser G, Wang W, Albrecht C, Rehahn M, Schwiegk S, Wegner G (1992) Polym Preprint. Am Chem Soc Div Polym Chem 33/1:294

6. Wang W, Lieser G, Wegner G (1993) Liq Cryst (in print)
7. Frenkel D (1988) J Phys Chem 92:3280
8. Eglinton G, McCrae W (1963) Adv Org Chem 4:225
9. Kanig G (1975) Colloid Polymer Sci 57:176
10. Voigt-Martin IG (1985) Adv in Polymer Sci 67:195
11. Wunderlich B (1973) Macromolecular Physics, vol 1, New York, London
12. Bassett DC (1981) Principles of Polymer Morphology, Cambridge Univ Press
13. Wegner G, Rodriguez-B. M, Lücke A, Lieser G (1980) Makromol Chem 181:1763
14. Folkes MJ, Keller A (1973) In: Haward RN (ed) The Physics of Glassy Polymers, London
15. Müller MA, Schmidt M, Wegner G (1984) Makromol Chem Rapid Commun 5:83
16. Raviso M, Aime JP, Fave JL, Schott M, Müller MA, Schmidt M, Baumgartl H, Wegner G (1988) J Phys France 49:861
17. Basedow AM, Ebert KH (1977) Adv Polym Sci 22:83
18. Elias HG (1981) Makromoleküle, Basel p 688
19. Hudson SD, Thomas EL (1987) Mol Cryst Liq Cryst 153:63
20. Mazelet G, Kléman M (1988) J Mater Sci 23:3055
21. Shiwaku T, Nakai A, Hasegawa H, Hashimoto T (1990) Macromolecules 23:1590
22. Ford JR, Bassett DC, Mitchell GR, Ryan TG (1990) Mol Cryst Liq Cryst 180B:233
23. Yase K, Schwiegk S, Lieser G, Wegner G (1992) Thin Solid Films 210/211:22
24. Northolt MG, Sikkema DJ (1990) Adv in Polym Sci 98:115

Authors' address:

Dr. G. Lieser
MPI für Polymerenforschung
Postfach 3148
55021 Mainz, FRG

# The ultimate toughness of polymers. The influence of network and microscopic structure

M. C. M. van der Sanden, H. E. H. Meijer, and P. J. Lemstra

Centre for Polymers and Composites, Eindhoven University of Technology, Eindhoven, The Netherlands

*Abstract:* The deformation and toughness of amorphous glassy polymers is discussed in terms of both the molecular network structure and the microscopic structure. Two model systems were taken into consideration: polystyrene-poly(2,6-dimethyl-1,4-phenylene ether) blends (PS-PPE) and epoxides based on diglycidyl ether of bisphenol A (DGEBA). The network structure of the thermoplastic PS-PPE system could be varied systematically by changing the relative volume fractions of PS (low entanglement density, $\nu_e = 3 \times 10^{25}$ chains $m^{-3}$) and PPE ($\nu_e = 13 \times 10^{25}$ chains $m^{-3}$) in this miscible blend. The crosslink density, $\nu_c$, of the DGEBA system could be set by selecting various epoxide monomer molecular weights ($8 \times 10^{25} \leqslant \nu_c \leqslant 235 \times 10^{25}$ chains $m^{-3}$). The microscopic structure at length scales of 50—300 nm was controlled by the addition of different amounts of non-adhering core-shell-rubber particles having a constant diameter. Toughness is mainly determined by the maximum macroscopic draw ratio since the yield stress of most polymers approximately is identical (50—80 MPa). It is shown, based on the analysis of experimental data published in literature, that the theoretical maximum draw ratio, derived from the maximum (entanglement or crosslink) network deformation, is obtained macroscopically when the characteristic length scale of the microstructure of the material is below a certain value, i.e., the critical matrix ligament thickness between added non-adhering rubbery particles ("holes"). The value of the critical matrix ligament thickness ($ID_c$) uniquely depends on the network structure: at an increasing network density, $ID_c$ increases independent of the nature of the network structure (entanglements or crosslinks). A simple model is presented, based on an energy criterion, to account for the phenomenon of a critical ligament thickness.

*Key words:* Toughness — network density — material critical thickness

## Introduction

The last two decades have seen impressive progress concerning the development of polymer materials approaching the ultimate properties in terms of modulus and strength [1—3]. These maximum properties were achieved upon aligning and extending the polymer chains. For flexible macromolecules (notably ultra-high molecular weight polyethylene, UHMW-PE), this was realized by stretching the disentangled polymer network to its maximum [2]. The maximum extension ratio is determined by the molecular weight between network loci of enhanced friction (molecular weight between entanglements, $M_e$ — or, analogously, for thermosets molecular weight between crosslinks, $M_c$ —). According to classical rubber elasticity theory [4], the maximum draw ratio of a network is proportional to:

$$\lambda_{\max} \sim M_e^{1/2} . \tag{1}$$

In order to achieve high draw ratios of UHMW-PE, $M_e$ was increased by crystallizing from dilute solutions. Subsequently, the ease of deformability of the — near single — PE crystals allows for a high draw ratio in the solid state, even after removal of the solvent, resulting in a nearly perfect parallel orientation and close to complete extension of the macromolecules. The success of this procedure is on one hand limited by the deformability of the crystals (e.g., polyamides cannot be stretched to high draw ratios since the hydrogen bonds present in the crystal cause premature fracture during solid-state drawing) while, on the other hand, for amorphous glassy polymers the diluted network in the solution cannot be preserved during solidification due to the process of "reentangling" of the macromolecules during evaporation of the solvent.

Recently, we focused our attention on the achievement of the ultimate toughness of polymeric materials based on the above understanding of the drawing behavior of entangled or crosslinked networks. It was postulated [7] that, since the yield stress of a great variety amorphous glassy polymers approximately is identical (50—80 MPa), toughness is determined mainly by the maximum elongation at break. The characteristics of the network topology are mainly determined by the intra-molecular structure of the polymer molecule (chain stiffness). Based on these simple arguments, polystyrene (PS) should be one of the more ductile amorphous polymers, since it possesses a molecular weight between entanglements, $M_e$, of 19.1 kg mole$^{-1}$, resulting in a theoretical maximum draw ratio of 4.2 (320% strain-at-break). Kramer et al. have determined the maximum draw ratio of various amorphous polymers on the microscopic level [5, 6]. They analyzed a large number of glassy polymers by measuring the local extension ratio of the material inside a craze-fibril or shearing zone, and found a close agreement between the theoretical value (according to Eq. (1)) and the strain measured microscopically.

Commonly, PS, however, is a very brittle polymer that exhibits a macroscopic strain-at-break typically of less than 5% to be compared with 320% as discussed above. The macroscopic strain-to-break is governed by a non-homogeneous deformation behavior (stress and strain localizations). Thus, in order to extend the high strain levels from a microscopic to the macroscopic level, the polymer microstructure has to be controlled so as to prevent catastrophic fracture anywhere in the sample. As a general rule: the stress in the locally initiated deformed regions should not surpass the tensile strength while the stress in the connected undeformed material should be higher than the yield stress.

Previously, we have shown that the macroscopic strain-to-break of PS can be increased from less than 5% up to 200% by introducing approximately 60 vol% of holes in the polymeric material [7]. The critical dimension of the polymeric material between the holes, below which the polymer behaves in a ductile manner, is called the critical ligament thickness and depends on the type of polymer. In addition, the value of the critical ligament thickness proves to be a well defined function of the entanglement or network density [8, 9].

In this paper, we present a generalized view of deformation and toughness of polymeric systems. The detailed experimental procedures are presented elsewhere in a series of previously published papers on this subject [7—9]. This paper intends to provide an overall insight into the principles of deformation and toughness of glassy, amorphous polymers based on the experimental evidence presented elsewhere [7—9]. Both, neat and rubber-modified thermoplastic and thermosetting polymers are taken into consideration with a focus on the macroscopic toughness (read: strain-at-break) in dependence of the molecular network structure and the microstructure (ligament thickness). The molecular weight between entanglements ($M_e$), as discussed previously, was varied by the compositions of the homogeneously miscible system polystyrene-poly(2,6-dimethyl-1,4-phenylene ether) (PS-PPE) [10]. For the thermosetting polymers the molecular weight between crosslinks was varied by changing the monomer molecular weight of the epoxide resin used [9]. Rather loose chemical networks were chosen in our investigations in order to obtain values of $M_e$ and $M_c$ of the same order of magnitude, in terms of the network density, thus elucidating the principle similarity in the deformation behavior of both types of networks. A non-adhering core-shelll-rubber of constant particle size was used in order to control the interparticle distance in the case of rubber-modified polymers. Toughness was determined via slow speed tensile testing as well as via high-speed notched tensile tests at various temperatures.

## Thermoplastics (physical networks)

The entanglement molecular weight ($M_e$) can be adjusted between the value of pure PS (19.1 kg mole$^{-1}$) and of pure PPE (3.4 kg mole$^{-1}$) by changing the relative volume fractions of both components in this miscible blend (see ref. [8]). The characteristics of the polymer network can be determined in the melt where the secondary bonds are excessively weak. By determining the rubber plateau modulus ($G_{No}$), using dynamic mechanical thermal analysis, the molecular weight between entanglements can be calculated using Eq. (2):

$$M_e = \frac{\rho RT}{G_{No}} , \qquad (2)$$

where $\rho$ is the density, $R$ is the gas constant and $T$ is the reference temperature [11]. The entanglement density ($\nu_e$) can be calculated with:

$$\nu_e = \frac{\rho N_A}{M_e} , \qquad (3)$$

where $N_A$ is Avogadro's number. This yields a value of $\nu_e$ of $3 \times 10^{25}$ chains $m^{-3}$ for pure PS and a value of $13 \times 10^{25}$ chains $m^{-3}$ for pure PPE. With Eq. (1), the theoretical maximum draw ratio, $\lambda_{max}$, can be calculated from the experimentally determined molecular weight between entanglements.

In order to compare the theoretical maximum draw ratio with the macroscopic draw ratio of PS-PPE blends the macroscopic strain-to-break of neat and core-shell-rubber modified PS-PPE blends is measured. In Fig. 1 the experimentally determined macroscopic strain-to-break data of various PS-PPE blends are shown as a function of the core-shell-rubber concentration present in the matrix, as obtained from slow speed tensile testing (5 mm $min^{-1}$) (see ref. [8]). The core-shell-rubber (particle size: 0.1-0.3 μm) is present to generate locally thin ligaments since the polymethylmetacrylate shell does not attach to the PS, respectively, the PS-PPE, matrix [12]. The macroscopic strain-to-break of pure PS is less than 5% (not incorporated in Fig. 1). Only if 60 wt% of the non-adhering core-shell-rubber is added to PS (curve A) the macroscopic strain-at-break is significantly increased ($\approx$ 200%). Scanning electron microscopy revealed that the rubbery particles still form the dispersed phase even in these highly filled systems [7].

The average matrix ligament thickness at which the brittle-to-ductile transition occurs for PS was calculated from the rubber volume fraction and particle size assuming a body-centered lattice. PS modified with 50 wt% core-shell-rubber contains

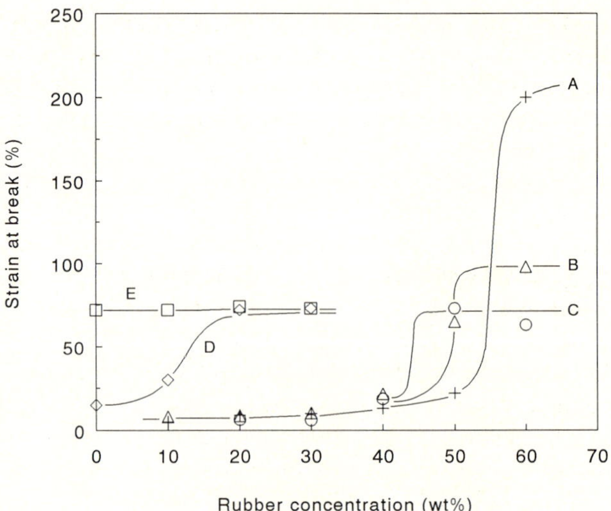


Fig. 1. Strain-at-break of core-shell-rubber modified PS-PPE blends vs rubber concentration with parameter the matrix composition (PS-PPE): A: 100—0; B: 80—20; C: 60—40; D: 40—60; and E: 20—80

ligaments with an average thickness of 0.06 μm and possesses a typically brittle fracture behavior, while PS modified with 60 wt% rubber (average ligament thickness: 0.04 μm) demonstrates a ductile deformation behavior. Hence, the critical matrix ligament thickness is about 0.05 μm for PS. If adhering core-shell-rubbers are used (e.g., with a polystyrene shell [7]), even blends containing 60 wt% rubber show a brittle fracture behavior, thus emphasizing that "holes" are locally needed. Obviously, particles that are attached to the matrix prevent the formation of isolated thin matrix ligaments that can deform without any interference of the dispersed rubbery phase.

If 20 wt% PPE is present in the matrix (PS-PPE 80—20 w-w; curve B) the brittle-to-ductile transition is found to be at 50 wt% rubber (average ligament thickness: 0.06 μm). Interestingly, the macroscopic strain-to-break in the ductile region does not increase any further if the matrix ligament thickness is already below its critical value. The maximum level of macroscopic strain-to-break is lower for PS-PPE 80—20 w-w compared with pure PS.

Upon increasing the PPE content of the matrix (curves C and D), the trends are confirmed as described for PS-PPE 80—20 w-w blends: increasing the PPE content results in i) a shift of the critical rubber concentration to lower values (hence, a shift of the critical matrix ligament thickness to higher values) and ii) a decrease of the maximum macroscopic strain-at-break of the rubber-modified PS-PPE blends.

For the PS-PPE 20—80 w-w blend (curve E), the maximum macroscopic strain-at-break is already obtained in the neat blend (no rubber present). Adding core-shell rubbery particles to this PS-PPE blend does not result in an increase in strain-to-break, therefore, this blend does not show a brittle-to-ductile transition. However, under more extreme testing conditions (high-speed, or notched tensile testing [13]) a brittle-to-ductile transition becomes noticeable, as expected.

## Thermosets (chemical networks)

Crosslinked epoxides, based on diglycidyl ether of bisphenol A (DGEBA), that are stoichiometrically cured with 4,4′-diamino diphenyl sulphone (DDS) possess a molecular weight between crosslinks ($M_c$) typically in the range of 0.3—7 kg mole$^{-1}$ (determined analogous to the value of $M_e$ using Eq. (2)). In terms of network density these $M_c$ values correspond to $8 \times 10^{25}$ — $235 \times 10^{25}$ chains m$^{-3}$, i.e., network densities ranging from values comparable with, up to values much higher than, the thermoplastic PS-PPE model system.

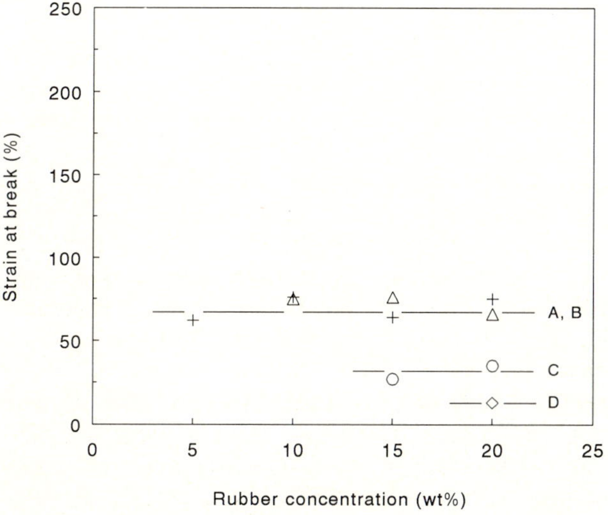


Fig. 2. Strain-at-break of core-shell-rubber modified epoxides vs rubber concentration with parameter the molecular weight between crosslinks: A: 6.8 kg mole$^{-1}$; B: 4.4 kg mole$^{-1}$; C: 1.6 kg mole$^{-1}$; and D: 0.9 kg mole$^{-1}$

In Fig. 2, the macroscopic strain-at-break data of various neat and core-shell-rubber modified crosslinked epoxides, measured in slow speed tensile testing (crosshead speed 5 mm min$^{-1}$), are shown (for experimental details see ref. [9]). Curves A and B correspond to epoxides with a rather high value of $M_c$ (curve A: $M_c$ = 6.8 kg mole$^{-1}$; curve B: $M_c$ = 4.4 kg mole$^{-1}$). The macroscopic strain-at-break of these epoxides is constant, independent of the amount of core-shell-rubber added to the system. The strain-to-break data of the pure epoxides are omitted in Fig. 2 because the reliability of these data is rather poor due to the high defect sensitivity of the samples.

Decreasing the molecular weight between crosslinks (Fig. 2 curve C, respectively curve D) results in a decrease of the macroscopic draw ratio (read: strain-to-break) analogous to the results obtained with the thermoplastic blends of PS and PPE where a decrease in $M_e$ also resulted in a lower value of the maximum macroscopic strain-at-break (see Fig. 1).

## Macroscopic draw ratio

In Fig. 3 the strain-to-break data of Figs. 1 and 2 are combined: the maximum macroscopic draw ratio, $\lambda_{macr}$, is plotted vs the network density. The network density is expressed in terms of respectively entanglement density (thermoplastic PS-PPE system) or crosslink density (thermosetting epoxide system).

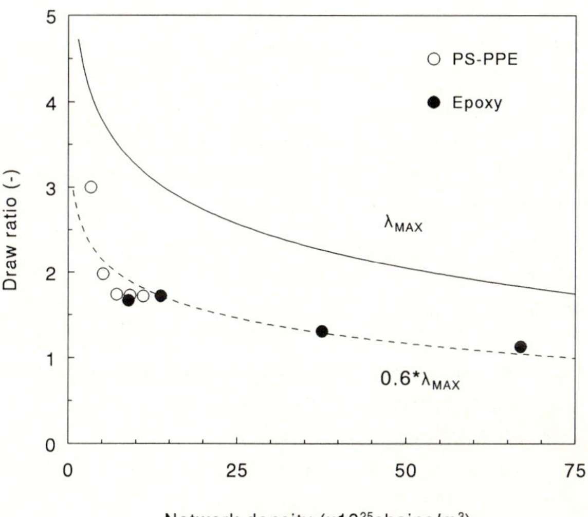


Fig. 3. Draw ratio vs network density (entanglement and/or crosslink). For details see text

As can be inferred from Fig. 3, the maximum macroscopic draw ratio decreases with a decreasing molecular weight between entanglements (for the system PS-PPE, open circles) or crosslinks (for the thermosetting system, filled circles). The drawn line in this figure is calculated according to the maximum draw ratio of a single network strand (Eq. (1)). The experimental data all are systematically situated below the drawn line.

Also experiments of Kramer et al. showed the same discrepancy [5, 6] as observed in Fig. 3. These authors measured the draw ratio inside a shear deformation zone and inside a craze fibril (i.e., microscopic level). Comparing these microscopic strain-to-break data with the theoretical maximum draw ratio of the various polymers shows, in the case of the crazing mechanism, a clear one-to-one agreement. For shear deforming polymers only 60% of the theoretical draw ratio is measured inside a shear deformation zone. The reason for this discrepancy is explained by the rather naive assumption of taking one single network strand in order to describe a fully three dimensional network [6]. The close agreement in the case of the crazing mechanism is only misleading because, in this situation, the network structure is partly destroyed during the formation of the craze structure. In Fig. 3 also the line according to $0.6 \cdot \lambda_{max}$ is drawn (dashed line). A good correlation is obtained between the experimental data and the drawn dashed line. Obviously, the distribution in network density limits the maximum macroscopic draw ratio to 60% of the theoretical draw ratio of the average network density. Any network arguments only lead to higher predicted values of the network extension ratio.

From Fig. 3 it is clear that the maximum macroscopic draw ratio is governed by the molecular network structure. The entanglement (or, alternatively, crosslink) network governs the deformation behavior independent of the nature of the network: physical network loci or chemical crosslinks. Data obtained from the thermoplastic PS-PPE system perfectly match with the data obtained from the epoxide system. Interestingly, there is even some overlap in both systems, which further corroborates the previous statement.

## Critical ligament thickness

As already mentioned during the discussion of Fig. 1, the maximum macroscopic strain-to-break in thermoplastic systems often can be obtained only if the material is filled with a certain concentration of "holes" (non-adhering rubbery particles). The ligaments between those holes need to have a thickness below the critical value in order to obtain the maximum value of macroscopic toughness. For the PS-PPE 20—80 w-w blend, however, no brittle-to-ductile transition was observed during uniaxial slow speed tensile testing, analogous to the thermosetting epoxy system. Fig. 4 shows that in the latter situation also brittle-to-ductile transitions are observed if the testing conditions are more extreme: e.g., notched samples strained at a crosshead speed of 1 m $s^{-1}$ at different temperatures [9].

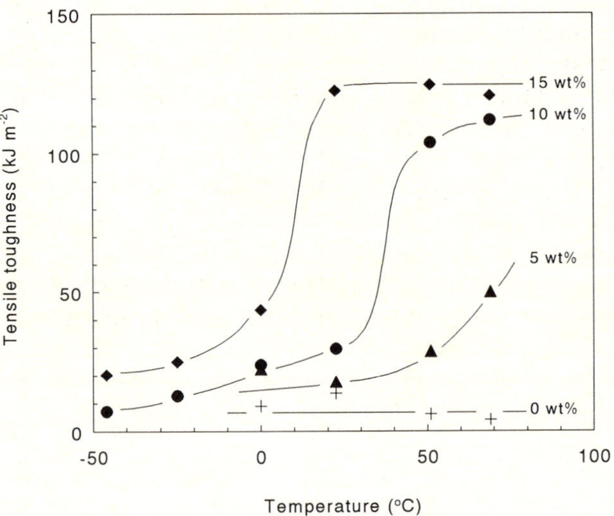


Fig. 4. Notched tensile toughness of core-shell-rubber modified epoxide ($M_c$ = 6.8 kg mole$^{-1}$) vs temperature with parameter the rubber concentration

In Fig. 4 the tensile toughness (integrated area under the recorded stress strain curve) is plotted vs temperature for a crosslinked epoxide of a $M_c$ value of 6.8 kg mole$^{-1}$ at various rubber concentrations. Pure epoxide (0 wt% rubber) exhibits brittle fracture within the investigated testing range. When 5 wt% core-shell-rubber is present the tensile toughness already shows the characteristics of a brittle-to-tough transition with increasing temperature. In the event that the rubber concentration is further increased (10 wt%) the transition temperature is lowered and the maximum attainable toughness is increased. Above the transition temperature the tensile toughness does not increase.

When 15 wt% core-shell-rubber is added to the epoxide, only the transition temperature is shifted to lower values while the maximum level of tensile toughness remains the same. As demonstrated by Borggreve [14] the brittle-to-tough transition temperature is correlated directly with the matrix ligament thickness between the easy cavitating rubbery particles. The transition temperature is found to be independent of the glass transition temperature of both the rubbery phase and the matrix.

A plot of the high-speed tensile toughness of the various epoxides, of different values of $M_c$, vs the rubber concentration (reference temperature 20 °C) visualizes the influence of $M_c$ on the position of the brittle-to-ductile transition (see Fig. 5). Curve A corresponds to the epoxide of a molecular weight between crosslinks of 6.8 kg mole$^{-1}$ of which the detailed data are shown in Fig. 4. In accordance with Borggreve's results with nylon-6, it can be inferred that a brittle-to-tough transition can be realized in these systems, not only by an increase in temperature, but also by a decrease of the local matrix ligament thickness. When the $M_c$ value of the epoxide is lowered (curve B; $M_c$ = 4.4 kg mole$^{-1}$), the brittle-to-ductile transition is found at a lower rubber concentration. Thus, the critical ligament thickness increases with a decreasing molecular weight between crosslinks.

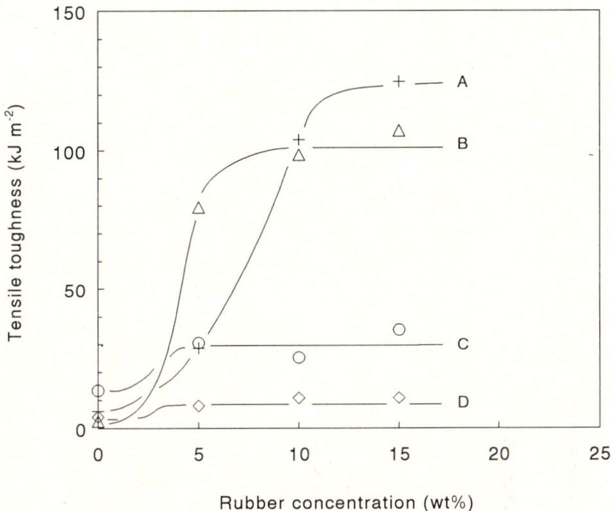


Fig. 5. Notched tensile toughness of core-shell-rubber modified epoxides vs rubber concentration with parameter the molecular weight between crosslinks: A: 6.8 kg mole$^{-1}$; B: 4.4 kg mole$^{-1}$; C: 1.6 kg mole$^{-1}$; and D: 0.9 kg mole$^{-1}$ (reference temperature: 20 °C)

Further decreasing the molecular weight between crosslinks results in a further increase of the critical interparticle distance (curve C, respectively, curve D of Fig. 5). For the epoxides possessing an $M_c$ of 1.6 kg mole$^{-1}$ (curve C) and 0.9 kg mole$^{-1}$ (curve D) the brittle-to-ductile transition is positioned below 5 wt% core-shell-rubber. Hence, the exact value of the critical interparticle distance cannot be accurately estimated for these two epoxides using the small ($\approx$0.2 μm) core-shell-rubbers. The maximum level of tensile toughness also depends on the value of $M_c$. In agreement with the conclusions arrived at in the discussion of Fig. 2, the maximum tensile toughness (read: macroscopic draw ratio) increases with an increasing molecular weight between crosslinks. This confirms the validity and usefulness of the concept of a maximum macroscopic draw ratio that is determined by the stretching of a network strand on the molecular level.

In Fig. 6 the values of the critical matrix ligament thickness, as obtained from slow speed tensile tests (PS-PPE system, open circles; Fig. 1) and obtained from high-speed tensile tests (epoxide system, filled circles; Fig. 5) are plotted vs the network density. Fig. 6 clearly demonstrates that the critical ligament thickness increases with a decreasing $M_c$. Surprisingly, the value of the critical matrix ligament thickness for the epoxide of an $M_c$ value of 6.8 kg mole$^{-1}$ ($\nu_c$ = 8.7 × 10$^{25}$ chains m$^{-3}$), as measured during a notched high-speed tensile test, agrees well with the value obtained from slow speed tensile testing of a PS-PPE 40—60 w-w blend ($\nu_e$ = 9 × 10$^{25}$ chains m$^{-3}$).

In conclusion, the network structure determines not only the maximum macroscopic draw ratio, but also uniquely dictates the value of the critical matrix ligament thickness, below which the maximum value of the macroscopic draw ratio is observed, and, moreover, the exact value of this critical thickness is rather insensitive to the testing conditions applied and the nature of the network structure (entanglements or crosslinks). By definition, the critical ligament thickness is the material specific thickness below which the polymer network can be stretched to its theoretical extension ratio, independent of the testing conditions applied. The validity of this statement will be demonstrated in a future paper [13] where the notched high-speed tensile toughness of the PS-PPE system will be extensively described and compared with the slow speed tensile data reported in this study.

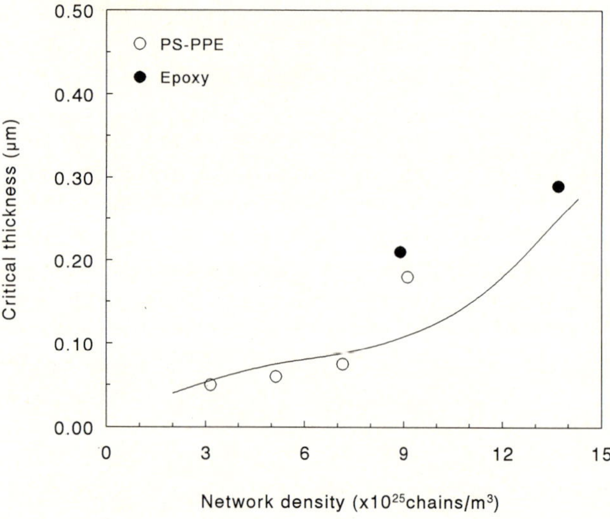

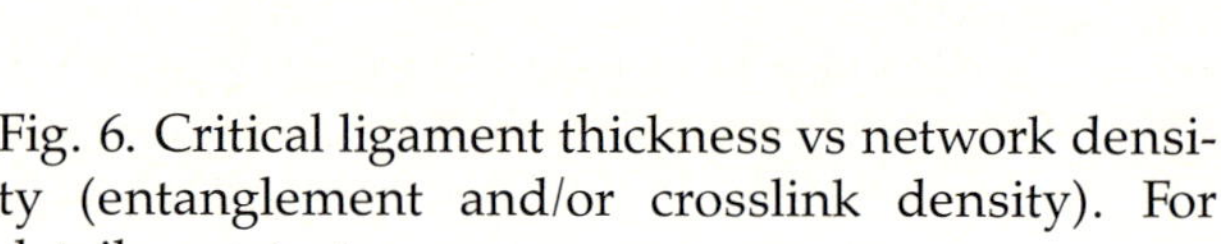


Fig. 6. Critical ligament thickness vs network density (entanglement and/or crosslink density). For details see text

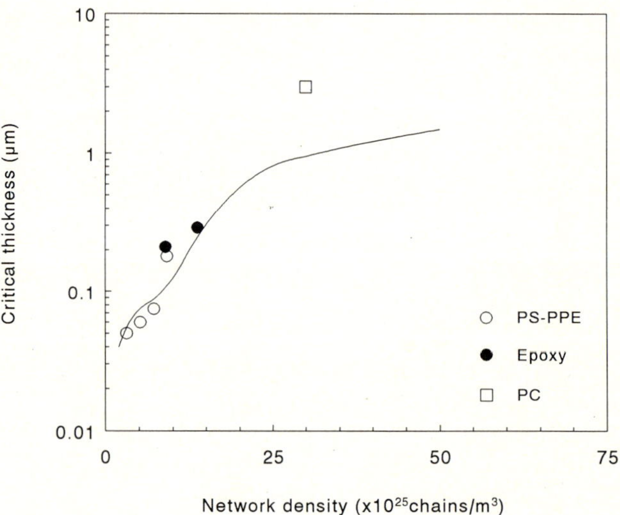


Fig. 7. Same as Fig. 6, but on a different scale

Fig. 7 shows the same data as Fig. 6 on a different scale, allowing for the incorporation of the critical ligament thickness of polycarbonate [15] (obtained from notched high-speed tensile tests). The comparison with the other data presented in Fig. 7 is of interest since polycarbonate represents one of the most densely entangled thermoplastic polymers known. More densely crosslinked epoxides can be easily obtained because the more practical systems possess crosslink densities far above $30 \times 10^{25}$ chains $m^{-3}$. In Fig. 7 no data for this class of epoxides are available yet since the exact value of the critical ligament thickness for low $M_c$ thermosets can only be determined accurately with large rubbery particles. In order to allow for a direct comparison, in this study a constant, rather small sized, rubbery filler was chosen yielding unpractical small volume fractions for the determination of the exact value of the critical thickness at low values of $M_e$ or $M_c$. The area of the right-hand side of Fig. 7 will be investigated in the near future.

## Model description

The physical origin of the existence of a critical matrix ligament thickness for the occurrence of the brittle-to-ductile transition can be qualitatively understood in terms of continuation of locally initiated deformation without premature fracture of deformed areas. The locally initiated deformation can continue throughout the complete sample if the stress in the deformed regions does not surpass the tensile strength while the stress in the connected undeformed material is higher than the yield stress. Hence, in order to derive a simple model describing the phenomenon of the material thickness dependent brittle-to-ductile transition, the conditions have to be investigated that allow for a continuation of local deformation without catastrophic failure. In Part 2 [8], a quantitative model was presented based on an energy criterion where, a comparison is made between the energy available (stored elastic energy) and the energy needed to create brittle fracture (fracture surface energy). This simple model will be recapitulated briefly. In Fig. 8 the model is schematically illustrated.

The rubbery particles (see Fig. 8a), represented by the two circles, generate a matrix ligament containing elastic energy, $U_{AV}$, proportional to the third power of $d_1$ (i.e., proportional to the shaded volume):

$$U_{av} = \frac{\pi}{6} (d_1)^3 \frac{\sigma_y^2}{2 E_1} , \qquad (4)$$

where $\sigma_y$ is the yield stress and $E_1$ is the Young's modulus of the matrix. The potential fracture surface is proportional to the surface of a circle of diameter $d_2$ (see Fig. 8b). The criterion for fracture is $d_2 = d_1/\sqrt{(\lambda_{max})}$ since this represents the smallest

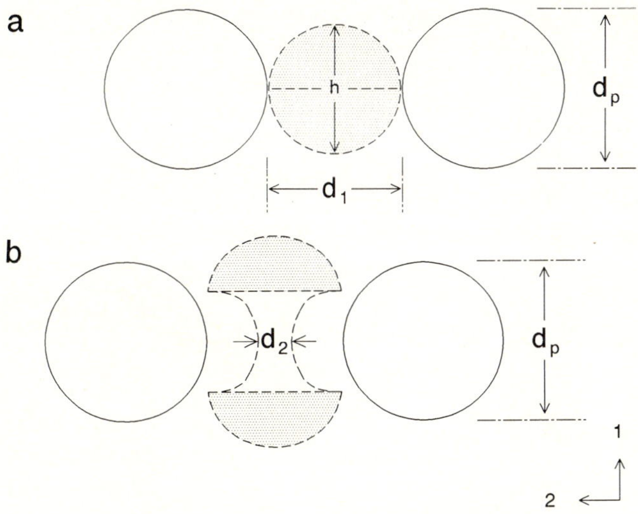


Fig. 8. Schematic view of a matrix ligament between two rubbery particles: a) before plastic deformation, and b) during plastic deformation

potential fracture surface and, therefore, the most critical situation with respect to brittle fracture. Hence, the energy, $U_{re}$, required to create a brittle fracture can be estimated to be:

$$U_{re} = 2\frac{\pi}{4}(d_2)^2\Gamma\,, \tag{5}$$

where $\Gamma$ is the surface energy of the polymeric matrix material. Equating the available elastic energy with the required surface energy results in an expression of the critical ligament thickness, $d_{1c}$, in dependence of several material parameters, i.e., network density, $\nu$ (entanglement and/or crosslink density), theoretical draw ratio $((d_1/d_2)^2)$, Young's modulus and yield stress (see ref. [8] for a more detailed description of the model):

$$d_{1c} = ID_c = \frac{6(\gamma + k_1\nu^{1/2})E_1}{\lambda_{max}\sigma_y^2}\,, \tag{6}$$

where $k_1$ is a constant ($k_1 = 7.13 \times 10^{-15}$ J chain$^{-1/2}$ m$^{-1/2}$, ref. [8]) and $\gamma$ is the Van der Waals surface energy.

The lines in Figs. 6 and 7 are drawn according to the simple model (Eq. (6)). Especially in the region of low network densities, the model predictions agree well with the experimental values of the critical ligament thickness. For higher network densities a slight deviation is observed between the predicted and experimental values of the critical ligament thickness. The latter is probably due to an overestimation of the stored elastic energy available to create a possible brittle fracture.

From Fig. 7 it is clear that the critical ligament thickness of densely crosslinked epoxides is very large; therefore, epoxides can be toughened to their maximum extent by adding only a small volume fraction of (preferably) non-adhering rubbery particles. However, the maximum attainable level of toughness of this class of materials is rather low compared to the low-entanglement density thermoplastic polymers (see Fig. 3).

## Conclusions

For both entangled thermoplastic polymers and crosslinked thermosets a maximum macroscopic draw ratio of about 60% of the (theoretical) natural draw ratio, $\lambda_{max}$, of these polymers can be achieved by control of the microstructure. Accordingly, the maximum macroscopic draw ratio (i.e., toughness) is found to scale with the square root of the molecular weight between network loci.

In order to obtain the high strain levels on a macroscopic scale, locally initiated deformation must continue without breakage of the deformed volumes anywhere in the sample. For polymers of a high molecular weight between network loci (e.g., PS), the sample needs to exhibit a small characteristic length scale ($\leqslant$0.05 μm), achieved by the introduction of non-adhering rubbery particles, in order to fulfil the conditions as described above. The value of the critical ligament thickness is a well defined function of the molecular network structure ($M_e$ or $M_c$), independent of the nature of the network structure and is rather insensitive to the testing conditions applied. The values of the measured critical ligament thickness vary from 0.05 μm for PS (slow speed tensile testing) to 0.3 μm for epoxide having a molecular weight between crosslinks of 4.4 kg mole$^{-1}$ (notched high-speed tensile testing). When the local thickness is further decreased below the critical level toughness will not increase any further. For epoxides having a molecular weight between crosslinks lower than 4 kg mole$^{-1}$ the critical ligament thickness is larger than 0.35 μm. As simple model was presented that quantitatively describes the existence of a critical ligament thickness and its dependence on the network density, theoretical draw ratio, Young's modulus and yield stress of the macromolecular material.

## References


1. Kwolek et al. (1977) Synthesis, anisotropic solutions, and fibers of poly(1,4-benzamide). Macromolecules 10:1390—1396
2. Smith P, Lemstra PJ, Booij HC (1981) Ultradrawing of High-Molecular-Weight Polyethylene cast from solution. II. Influence of initial polymer concentration. J Polym Sci Polym Phys Ed 19:877—888
3. Ward IM (1987) (ed) Developments in Oriented Polymers-2. New York, NY: Elsevier Applied Science Publ
4. Treloar LRG (1975) The physics of rubber elasticity. Oxford: Clarendon
5. Donald AM, Kramer EJ (1982) Deformation zones and entanglements in glassy polymers. Polymer 23:1183—1188
6. Kausch HH (1990) (ed) Advances in Polymer Science, vol 91/92. Berlin: Springer 1
7. van der Sanden MCM, Meijer HEH, Lemstra PJ (1993) Deformation and toughness of polymeric systems: 1. The concept of a critical thickness. Polymer 34:2148—2154
8. van der Sanden MCM, Meijer HEH, Tervoort TA (1993) Deformation and toughness of polymeric systems: 2. Influence of entanglement density. Polymer (in press)
9. van der Sanden MCM et al, Deformation and toughness of polymeric systems: 3. Influence of crosslink density. Polymer (accepted for publication)
10. Paul DR, Newman S (1978) (eds) Polymer Blends, Vol 2, Academic Press, San Diego, p 186
11. Ferry JD (1980) Viscoelastic Properties of Polymers. New York, John Wiley
12. Wu S (1982) Polymer interface and adhesion. Basel, Dekker
13. van der Sanden MCM et al., Deformation and toughness of polymeric systems: 4. Influence of strain-rate and temperature. Polymer (submitted)
14. Borggreve RJM (1988) Toughening of Polyamide-6. Ph D thesis, Twente University of Technology, 34—53
15. Wu S (1990) Chain structure, phase morphology, and toughness relationships in polymers and blends. Polym Eng Sci 30:753—761


Authors' address:

M. C. M. van der Sanden
Centre for Polymers and Composites
Eindhoven University of Technology
P.O. Box 513
5600 MB Eindhoven, The Netherlands

# Synthesis, characterization and relaxation of highly organized side-chain liquid crystalline polymers

U. W. Gedde, H. Andersson, C. Hellermark, H. Jonsson, F. Sahlén, and A. Hult

Department of Polymer Technology, Royal Institute of Technology, Stockholm, Sweden

*Abstract:* This paper is a review of recent work performed in our department on highly oriented side-chain liquid crystalline polymers made from aligned mesomorphic vinyl ether monomers. The chemistry involved in making the polymers, the techniques used for obtaining alignment of the monomers, the physical structure of the oriented polymers, and the dielectric relaxation are the presented topics. Thermal stabilization by crosslinking using bifuctional vinyl ether monomers is also described. Crosslinked polymers exhibited an almost reversible degree of mesogen group orientation after heating to 200 °C.

*Key words:* Side-chain LCP — polymerization — orientation — structure — relaxation

## Introduction

This paper reviews our work on organized side-chain liquid crystalline poly(vinyl ether)s which are made by polymerizing aligned mesomorphic monomers. The fundamentals of the chemistry involved in making the polymers, the physical structure of the formed polymers, and the dynamics of the polymer chain as revealed by dielectric measurements are the topics. Included is also a section on thermal stabilization by crosslinking of the poly(vinyl ether)s. The repeating unit structures of the polymers dealt with in this review are presented in Fig. 1. The work is viewed in the perspective of previous and parallel work on side-chain liquid crystalline polymers.

In liquid crystalline (LC) polymers the mesogenic unit can be incorporated in the main chain or as a side group. Before 1978, the side-chain LC polymer was merely a backbone with linked mesogenic units although, since the mesogens were directly linked or spaced by just one or two atoms, the main-chain's tendency to form a random coil conformation disturbed the mesogen's tendency to arrange anisotropically. As a result many attempts to synthesise side-chain LC polymers failed and, if a mesogenic polymer was obtained, it usually possessed a very narrow temperature region in which the mesophase was stable and the transition temperatures were generally very high. In 1978, Ringsdorf et al. [1, 2] suggested that a decoupling of the main chain's and the mesogen's motions was possible through the insertion of a flexible spacer. After this breakthrough, an abundance of side-chain polymers have been synthesized and the combinations of main chains, spacers, and mesogens seem to be infinite. The typical way to synthesize this class of compounds is by polymerization of monomers resembling the repeating unit, i.e., a monomer with mesogen, spacer, and polymerizable group, although other routes exist, e.g., attachment of spacer and mesogen to a polymer backbone via reactions of functional groups [3—5].

The transformation of a mesogenic monomer to a polymer stabilizes the mesophase; a nematic monomer usually results in a smectic polymer and if the monomer is isotropic (but contains potential mesogens) a nematic polymer will be obtained [6]. At the same time, the temperature region of the mesophase also broadens considerably. The thermal transition temperatures increase with increasing molar mass up to a critical molar mass value above which it levels off. An increase in the flexibility of the backbone results in a decrease in the glass transition temperature and usually also in an increase in the isotropization temperature. For most

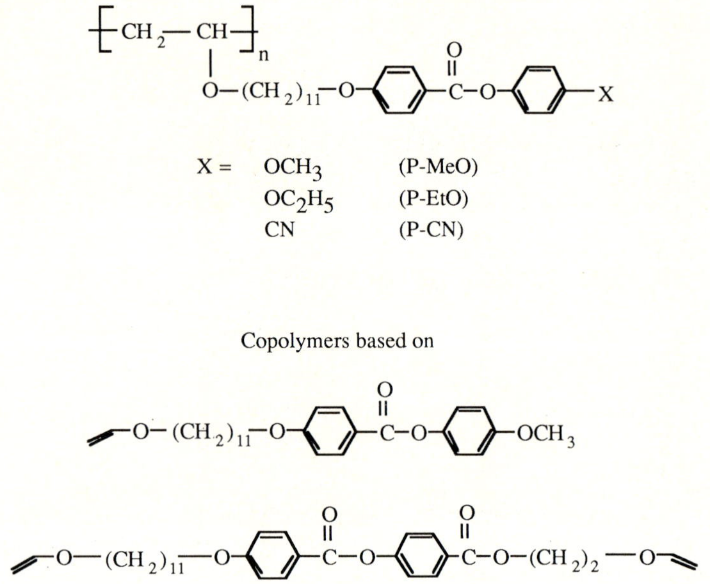

| X = | | |
|---|---|---|
| | $OCH_3$ | (P-MeO) |
| | $OC_2H_5$ | (P-EtO) |
| | CN | (P-CN) |

Copolymers based on

Fig. 1. Structure of poly(vinyl ether)s reviewed in this paper

side-chain LC polymers synthesized, complete decoupling between mesogen and backbone is not achieved [3]. Typical backbones used are poly(acrylate)s, poly(methacrylate)s, and poly(siloxane)s.

The spacer groups have a plasticizing effect on the backbone which results in a decrease in glass transition temperature with increasing spacer length. A longer spacer also increases the tendency for the formation of a smetic polymer [3]. The longer spacer group decouples the mesogenic group, building up the smectic layers more efficiently from the backbone and allows the former to attain higher perfection which is reflected in higher enthalpy, entropy, and temperature of isotropization [7]. By far the most used spacers are oligomethylenes, but oligoethylene oxide and oligosiloxane have also been used.

## Polymerization in thermotropic liquid crystalline media

An important question is whether the polymerization rate is affected by the order of the monomeric liquid. Perplies et al. reported [8] that the thermally initiated polymerization of Schiff-based acrylates and methacrylates resulted in a rate decrease when the reaction was carried out in the nematic instead of the isotropic state. Mesophase polymerization of styrene monomers carrying Schiff base moieties gave, on the other hand, no change of rate compared to polymerization in the isotropic phase, according to Paleos and Labes [9]. In contrast to these results, Hoyle et al. [10, 11] reported a rate enhancement in the LC phase upon photo-polymerization of cholesterol containing acrylate and methacrylate monomers. The polymerization rates in the smectic and cholesteric phases were very rapid compared to the rate in the isotropic state, presumably due to that alignment of monomers in the mesophase increases their accessibility for polymerisation. Furthermore, the rate in the cholesteric phase was higher than the rate in the smectic phase, and this was assigned to the higher translational mobility of the monomer in the former phase, the requirement for mobility overruling that for alignment.

Broer et al. [12—15] were the first to report in-situ photo-polymerization of mesomorphic, globally oriented monomers. The monomers, mono- and bifunctional acrylates, were oriented in their mesophase by conventional ordering techniques used for low molar mass liquid crystals, and then polymerised with uv-radiation with subsequent freezing-in of the structure. Oriented thin films were produced with a higher degree of order than films resulting from the conventional ordering of polymers.

The work which is reviewed in this paper, refs. [16—24], is based on polymers produced by photochemically- or thermally-induced cationic bulk polymerization of vinyl ether monomers. One advantage of photo-initiated over thermally initiated polymerization is the ease with which the polymerization temperature can be selected and the onset of polymerization can be directed, a very important fact when, for example, the polymerization is to be carried out in a monomer LC phase which is stable only within a very narrow temperature range. In order to perform controlled isothermal polymerizations with thermal initiation, the initiator has to be selected with caution. Another advantage of the photo-polymerization technique is that it is possible to change the polymerization rate by merely changing the intensity of the light. Usually, photopolymerization proceeds very fast, which suppresses phase segregation which is valuable in those cases when the formed polymer and its monomer are not compatible. All these factors are important if the polymerization process is used in the production of oriented polymeric structures, e.g., thin films.

The initiator systems used were based on onium salts of the photoredox type, i.e., the spectral response of the onium salt, diphenyliodonium hexafluorophosphate (**2**) and phenacyltetramethylenesulfonium hexafluoroantimonate (**4**), was broadened by either a free radical photo-initiator, 2,4,6-trimethylbenzoylethoxyphenylphosphine oxide (**1**) or a photosensitizer, phenothiazine (**3**) (Fig. 2).

Figure 3 shows the overall process for the production of the cationic initiating species. First, electron-donating free radicals are generated by photolysis or by excitation of a photosensitizer. Secondly, an electron is transferred from these species to the onium salt leading to the formation of the corresponding cation or radical cation, which in turn can initiate polymerisation. The initiator system **1** + **2** possesses a much higher reactivity than **3** + **4**, primarily due to the higher oxidising capability of the diaryliodonium salt (**2**) compared with the phenacylsulfonium (**4**).

The polymerization's can be carried out at different temperatures, i.e., from different physical states of the monomer liquid. 11-[p-(4-methoxyphenylbenzoate)oxy]undecanyl vinyl ether (*Meo*), 11-[p-(4-ethoxyphenylbenzoate)oxyundecanyl vinyl ether (EtO), and 11-[p-(4-cyanooxyphenylbenzoate)oxy]undecanyl vinyl ether (**CN**) polymerized from the isotropic, nematic, and smectic ($s_A$) states in a photo DSC [16] and between rubbed polyimide films to allow alignment of the monomer prior to

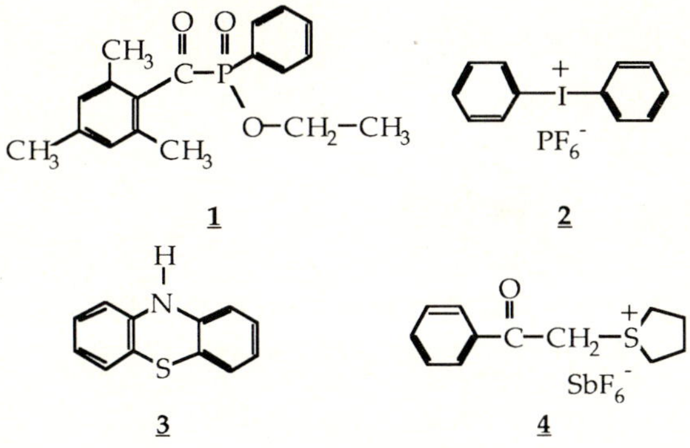

Fig. 2. Cationic photo-redox initiating systems

polymerisation and the results from characterization of the resulting polymers are presented in Table 1. The initial state of the monomer, nematic or isotropic, did not affect the molecular mass, molecular mass distribution or polymerization rate [16—19, 22]. It is clearly remarkable that molar masses of the polymers were high even at high temperatures of polymerization. Other characteristics are narrowness of molar mass distribution and high degree of conversion. The relatively low molar mass of polymers formed in contact with rubbed polyimide is most probably due to impurities leading to more extensive termination of polymerization [16].

The $2\theta$ positions of the first-order low-angle reflections of the 11-spacer polymers were equivalent with layer thicknesses between 3.0 and 3.5 nm, a fact which indicates that monolayers are the dominant feature [18—20]. P-EtO and P-MeO always exhibited untilted structures, $s_A$ at higher temperatures and crystal B structure at lower temperatures [19, 20]. P-CN exhibited a quite different behavior [20]. At higher temperatures a $s_A$

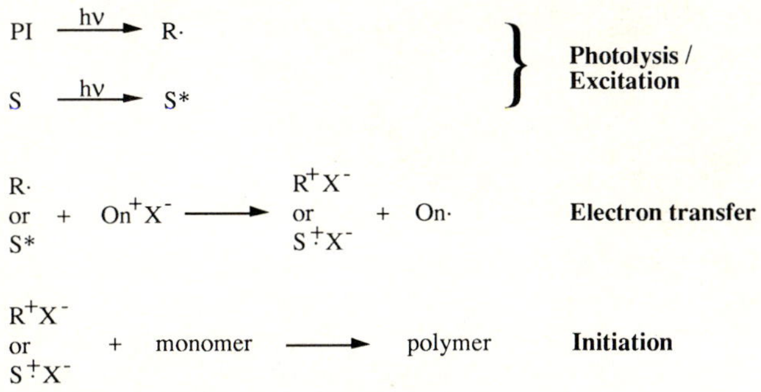

Fig. 3. Initiation of cationic polymerization through photoinitiated electron transfer reduction of onium salts. Abbreviations: PI — photo-initiator; S — photo-sensitizer; $On^+X^-$ — onium salt

structure similar to that of P-EtO and P-MeO was found. However, at temperatures below the low temperature first-order transition, a tilted smectic ($s_C$) phase was found. It may be suggested that the potentially strong interactions between the cyano end groups and adjacent phenyl groups "locks" the structure in a tilted position and that further ordering of the mesogens are prohibited. The inter-chain spacing amounted to 0.42—0.44 nm for all the studied polymers [18—20, 22].

The orientation data presented in Table 2 all refer to P-MeO and P-EtO samples of a thickness of 25 ± 5 μm polymerized between rubbed polyimide films with planar orientation of the mesogens. The measurements were performed at room temperature with the studied polymers in a crystal B state [20]. The orientation of the smectic layers was quite high, $f$ being approximately 0.95 [18, 19]. The wide-angle inter-chain reflection yielded a slightly lower degree of orientation ($f$ being between 0.90—0.95 for polymers produced from aligned nematic or smectic monomer) which points to the fact that the mesogens within the smectic layers exhibit some limited disorder. Polymerization of P-MeO at different temperatures under the influence of a surface field (rubbed polyimide film) yielded the following $f$ values based on the inter-mesogen chain spacing [19]: 80 °C (from isotropic monomer): 0.66; 54 °C (from nematic monomer): 0.93; 41 °C (from smectic $s_A$ monomer): 0.94. P-CN possessed a $s_C$ structure at room temperature and thus the order parameter based on infra-red dichroism of the

Table 2. Orientation, expressed in Hermans orientation function, of photo-polymerized aligned (rubbed polyimide film) nematic or smectic EtO and MeO

| | |
|---|---|
| Low-angle reflection (smectic layer orientation) | $f \approx 0.95$ |
| Wide-angle reflection (inter-chain spacing) | $f \approx 0.90$ |
| IR dichroism (spacer) | $f \approx 0.35$ |
| Dielectric spectroscopy | $\frac{\Delta\sigma_{\text{oriented}}}{\Delta\sigma_{\text{isotropic}}} \approx 1.5$ ($f \approx 1$) |

cyano stretching band (2250 $cm^{-1}$) amounted to 0.70, which is lower than in the case of P-EtO and P-MeO due to the higher degree of local order of the latter [24]. The spacer group orientation of this particular polymer was also measured from the C-H stretching bands and typical order parameter values were 0.3—0.4 [24].

## Dielectric relaxation (ref. [21])

Dielectric permittivity and loss have been measured over the frequency range of $10^{-3}$ Hz—1 kHz between 100 K and 350 K for samples of two mesomorphic side-chain poly(vinyl ether)s (P-EtO and P-CN; structures shown in Fig. 1) of different degree of mesogen group orientation and order. P-EtO was in a crystal B state at temperatures lower than 72 °C, whereas P-CN exhibited a smectic C phase at temperatures lower than 42 °C. Four

Table 1. Molecular and morphological structure of polyvinylethers (P-Meo, P-EtO and P-CN)

| | | | |
|---|---|---|---|
| $\bar{M}_n$ [a]) (g $mol^{-1}$) | 30—80 000 (thermal init.) | 20—40 000 (photo-init., unorient.) | 12—14 000 (photo-init., orient.) |
| $\frac{\bar{M}_w}{\bar{M}_n}$ | | 2.0 ± 0.3 | |
| Tactiticty [b]) | | atactic, $\frac{\text{meso}}{\text{racemo}} \approx 1$ | |
| Morphology [c]) | | Monolayers (for untilted structures 3—3.5 nm) | |
| Phases, transitions [d]) | | P-EtO; P-MeO:<br>P-CN: | $C_B^e - s_A - i$<br>$(g)\ s_C - s_A - i$ |

[a]) By SEC in PS equivalents [16—19, 22].
[b]) By C-13 NMR [19].
[c]) By X-ray diffraction [18—20].
[d]) LC structures were identified using x-ray diffraction [20].
[e]) $C_B$ should read crystal B [20].

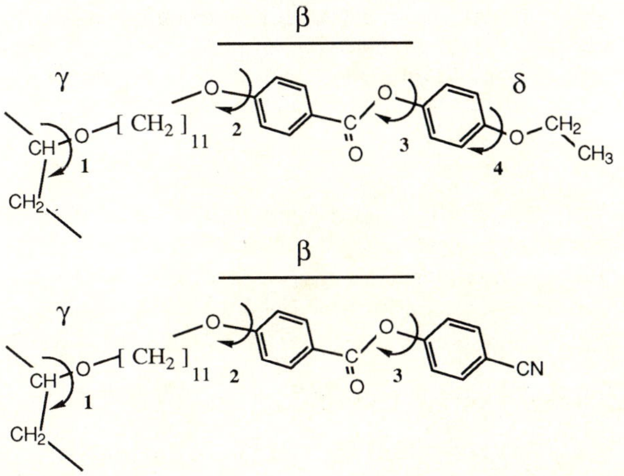

Fig. 4. Molecular assignment of dielectric relaxation of P-EtO and P-CN

relaxation transitions were found in P-EtO: α, the glass-rubber transition occurring at 290—300 K, and three subglass processes denoted β, γ and δ adapting Arrhenius temperature dependence with activation energies 74 to 108 kJ $mol^{-1}$ (β) and 33 kJ $mol^{-1}$ (γ). The activation energy and the relaxation strength of the high temperature subglass β process exhibited pronounced morphological and orientational dependence. The activation energy of the γ process was insensitive to morphological factors. The relaxation strengths of β and γ for the homogeneously oriented P-EtO sample were almost 50% higher than those of isotropic P-EtO, indicating very high orientation of the relaxing dipoles. P-CN exhibited three relaxation processes. α, β and γ. The low temperature process δ, only present in P-EtO, can be assigned to rotation of ethoxy end group. A complete chart with molecular assignment of the relaxation processes is shown in Fig. 4.

## Ordered and thermally stable films (ref. [19, 23])

The purpose of introducing bifunctional vinyl ether monomers (Fig. 5) was to prepare highly oriented side-chain liquid crystalline polymer film with a thermally stable mesophase. Compatibility of the monomers is an obvious requirement and mixtures of (**5**) with (**6**) and (**7**) respectively were not useful because both bifunctional monomers crystallized on cooling of the isotropic monomer mixtures, before any mesophase was formed.

Monomers (**5**) and (**8**) mixed intimely and did not phase separate upon cooling to the mesophase. The miscibility of monomers (**5**) and (**8**) can be ascribed to the structural similarity of the monomers. Polymerization of the monomer mixture (**5**) + (**8**) at 50 °C with photochemical initiation in a 30-μm-thick glass cell coated with a rubbed poly(imide) film

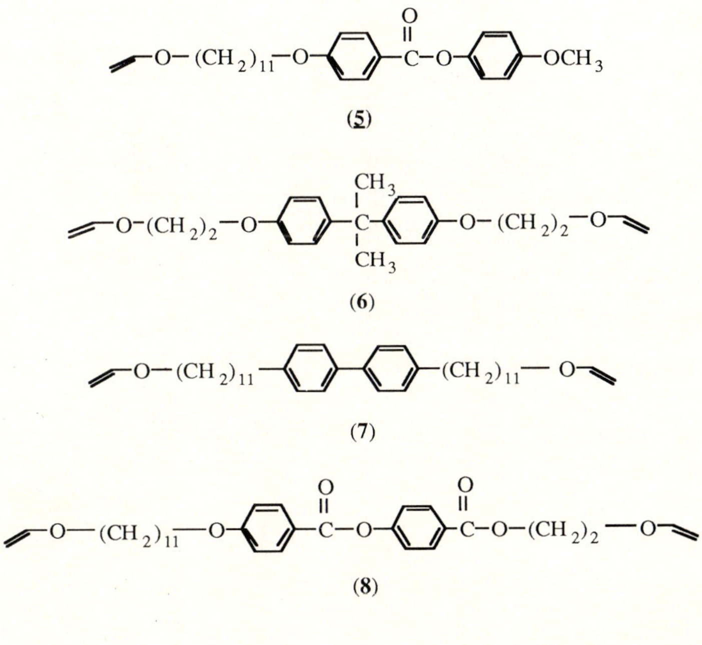

(**5**)

(**6**)

(**7**)

(**8**)

Fig. 5. Monomers used for preparation of thermally stable LCP

resulted in a highly ordered polymer film that could be heated up to 200 °C with largely preserved orientation after returning to room temperature. The film was completely transparent at room temperature and x-ray diffraction confirmed that structure of the crosslinked polymer was smectic *A*. A later IR dichroism study [24] showed that a minor part of the orientation was lost during the first heating scan up to 200 °C. The orientation remained however essentially reversible during further temperature cycling.

*Acknowledgements*

The reported studies have been sponsored by The National Swedish Board for Technical and Industrial Development (NUTEK) grant 86-03476P, The Swedish Natural Science Research Council (NFR) grants K-KU 1910-300, the Royal Institute of Technology (LUFT), AB Wilhelm Becker and Axel & Margeret Ax:son Johnsons' foundation. The collaborations within the programme with Drs. Richard Boyd and Fogou Liu, University of Utah, USA, Dr. Virgil Percec, Case Western Reserve University, USA and Drs. Javier Martinez Salazar and Fernando Ania, CSIC, Madrid, Spain are greatfully acknowledged.

**References**

1. Finkelmann H, Happ M, Portugall M, Ringsdorf H (1978) Makromol Chem 179:2541
2. Finkelmann H, Ringsdorf H, Siol W, Wendorff JH (1978) In: Blumstein A (ed) Mesomorphic Order in Polymers and Polymerization in Liquid Crystalline Media. ACS Symp Ser 74, American Chemical Society, Washington, DC
3. Percec V, Pugh C (1989) In: McArdle CB (ed) Side Chain Liquid Crystal Polymers. Chapman and Hall: New York
4. Platé NA, Freizon YS, Shibaev VP (1985) Pure & Appl Chem 57:1715
5. Keller P (1987) Macromolecules 20:462
6. Mark HF, Bikales NM, Overberger CG, Menges G (1987) Encyclopedia of Polymer Science and Engineering, 2nd edition, vol. 9, Wiley, New York
7. Gedde UW, Jonsson H, Hult A, Percec V (1992) Polymer 33:4352
8. Perplies E, Ringsdorf H, Wendorff JH (1974) Makromol Chem 175:553
9. Paleos CM, Labes MM (1970) Mol Cryst Liq Cryst 11:385
10. Hoyle CE, Chawla CP, Griffin AC (1988) Mol Cryst Liq Cryst Inc Nonlin Opt 157:639
11. Hoyle CE, Chawla CP, Griffin AC (1989) Polymer 60:1909
12. Broer DJ, Finkelmann H, Kondo K (1988) Makromol Chem 189:185
13. Broer DJ, Mol GN, Challa G (1989) ibid 190:19
14. Broer DJ, Boven J, Mol GN, Challa G (1989) ibid 190:2255
15. Broer DJ, Hikmet RAM, Challa G (1989% ibid 190:3201
16. Jonsson H, Sundell P-A, Percec V, Gedde UW, Hult A (1991) Polym Bull 25:649
18. Jonsson H, Andersson H, Sundell P-A, Gedde UW, Hult A (1991) Polym Bull 25:641
19. Andersson H, Gedde UW, Hult A (1992) Polymer 33:4014
20. Sahlén F, Ania F, Martinez-Salazar J, Hult A, Gedde UW (1993) Polymer (submitted)
21. Gedde UW, Liu F, Sahlén F, Hult A, Boyd RH (1993) Polymer (submitted)
22. Hellermark C, Gedde UW, Hult A (1992) Polym Bull 28:267
23. Andersson H, Gedde UW, Hult A (1993) Mol Cryst Liq Cryst (in press)
24. Sahlén F, Andersson H, Ania F, Hult A, Gedde UW (1993) Polymer (submitted)

Authors' address:

Dr. U. W. Gedde
Dept. of Polymer Technology
Royal Institute of Technology
S-10044 Stockholm, Sweden

# Author Index

Albrecht C 111
Andersson H 129

Bassett DC 23

Freedman AM 23

Gedde UW 129

Heise B 60
Hellermark C 129
Hult A 129

Jasse B 8
Jonsson H 129

Keller A 81
Kilian HG 60
Knechtel W 60
Kolnaar JWH 81

Lemstra PJ 120
Lieser G 111
Lotz B 32

Marikhin VA 39
Meijer HEH 120
Monnerie L 8
Myasnikova LP 39

Pertsev NA 52

Sahlén F 129
Smith P 32

Tassin JF 8

van der Sanden MCM 120

Ward IM 103
Wegner G 111
Wittmann JC 32

Ziabicki A 1
Zrinyi M 60

# Subject Index


chain extension 81
conformational defects 52
correlation of lamella thickness with $M_n$ 111

deformation 60
die drawing process 103
disclinations 52
dislocations 52
drawing 1

epitaxy 32
extension of connected networks 81
extremum principles 60

fibers 1
free flowing melts 81

high modulus 103
high strength flexible polymers 103

infrared dichroism 8
irradiation damage 111
irreversible thermodynamics 60

kink band 52

lamellae 23

main chain LCP 111
material critical thickness 120
molecular orientation 1
morphology 23

network density 120

orientation 8, 32, 60, 81, 129
orientation-stress characteristics 1

P-4-BCMU 11
plastic deformation 52
polyethylene 23, 60, 103
polymer blends 8
polymeric substrates 32
polymerization 129
polymers glasses 60
polystyrene 8
PTFE 32

relaxation 8, 129
rotational diffusion coefficient 1

side-chain LCP 129
spherulites 23
spinning 1
strain 1
— rate 1
stress 1
structure 129

toughness 120

uniaxial drawing 23